AF715558

FRANK HARGRAVES

The Universe's Final Secrets

Cosmic Endings from Big Crunch to Heat Death and Why Our Tiny Lives Matter

First published by Frank Hargraves 2025

Copyright © 2025 by Frank Hargraves

All rights reserved. No part of this publication may be reproduced, stored or transmitted in any form or by any means, electronic, mechanical, photocopying, recording, scanning, or otherwise without written permission from the publisher. It is illegal to copy this book, post it to a website, or distribute it by any other means without permission.

Frank Hargraves has no responsibility for the persistence or accuracy of URLs for external or third-party Internet Websites referred to in this publication and does not guarantee that any content on such Websites is, or will remain, accurate or appropriate.

Designations used by companies to distinguish their products are often claimed as trademarks. All brand names and product names used in this book and on its cover are trade names, service marks, trademarks and registered trademarks of their respective owners. The publishers and the book are not associated with any product or vendor mentioned in this book. None of the companies referenced within the book have endorsed the book.

This book is intended for informational purposes only. The author and publisher make no representations or warranties with respect to the accuracy or completeness of the contents of this work. The ideas, theories, and speculations expressed are those of the author and do not necessarily reflect the views of any institutions or individuals mentioned. The publisher is not liable for any damages or consequences arising from the use of this material.

First edition

This book was professionally typeset on Reedsy.
Find out more at reedsy.com

Contents

Introduction

The Cosmic Question...

What is the fate of the universe? It's a question that has echoed through the ages, whispered by philosophers, pondered by scientists, and dreamed of by poets. It's a question that transcends the boundaries of time and space, reaching into the very heart of existence. To ask about the fate of the universe is to ask about the ultimate destiny of everything we know—every star, every planet, every atom, and every fleeting moment of consciousness. It's a question that forces us to confront the limits of our understanding and the vastness of the unknown.

Imagine, for a moment, standing on the edge of a cliff, gazing out into an endless expanse. Below you lies an ocean, not of water, but of stars—billions upon billions of them, swirling in galaxies that stretch as far as the eye can see. This is the universe, a cosmic tapestry woven from the threads of time, space, and energy. But what lies beyond the horizon? Where is this vast, unfathomable expanse headed? Will it continue to expand forever, fading into cold, dark emptiness? Or will it collapse back in on itself, ending in a fiery cataclysm? Or perhaps, will it tear itself apart in a final, violent act of cosmic destruction?

These are not just abstract musings—they are the questions that drive modern cosmology. They are the questions that have led us to some of the most profound discoveries in human history, from the Big Bang to dark energy, from the cosmic microwave background to the accelerating expansion of the universe. They are the questions that remind us of our place in the cosmos: tiny, fleeting, and yet infinitely curious.

This book is a journey into the unknown, a voyage through the cosmos to explore the most profound theories about the fate of the universe. It's a journey that begins with the fiery birth of the universe in the Big Bang and takes us through the eons, past the formation of galaxies, stars, and planets, to the very edge of reality itself. Along the way, we'll encounter some of the most mind-bending ideas in science: the Big Crunch, the Big Freeze, the Big Rip, and even the possibility of a multiverse. We'll delve into the science behind these theories, exploring the evidence, the models, and the predictions that shape our understanding of the cosmos.

But this is not just a book about science. It's a book about wonder, about the thrill of discovery, and about the profound questions that arise when we contemplate the ultimate destiny of the universe. What does it mean for the universe to end? What does it mean for us, as conscious beings, to live in a universe that is finite, that had a beginning, and that will one day have an end? These are questions that touch on the very nature of existence, on the meaning of life, and on the limits of human knowledge.

As we embark on this journey, we'll encounter some of the greatest minds in history—scientists like Albert Einstein, Edwin Hubble, Stephen Hawking, and countless others who have dedicated their lives to unraveling the mysteries of the cosmos. We'll explore the tools and technologies that have allowed us to peer into the depths of space and time, from the Hubble Space Telescope to the Large Hadron Collider. And we'll grapple with the profound implications of what we've learned, from the nature of dark energy to the possibility of other universes.

But this journey is not just for scientists or philosophers. It's for anyone who has ever looked up at the night sky and wondered what lies beyond. It's for anyone who has ever felt the thrill of discovery, the awe of the unknown, and the profound sense of connection that comes from realizing that we are all part of something much larger than ourselves.

So, as we stand on the edge of this cosmic cliff, gazing out into the infinite expanse, let us take a deep breath and step forward. The journey ahead is vast, filled with mysteries and wonders beyond imagination. But it is a journey worth taking, for it is a journey into the very heart of existence—a journey to discover the fate of the universe, and perhaps, in the process, to discover something about ourselves.

Welcome to the adventure. Welcome to the edge of reality. Welcome to the quest to uncover the ultimate destiny of the cosmos.

Chapter 1: The Big Bang and the Expanding Universe

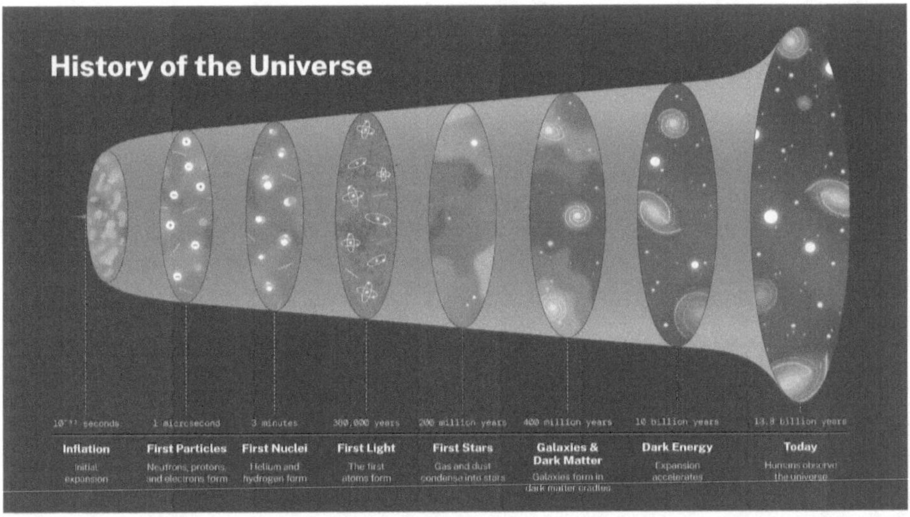

NASA's Universe Evolution Timeline (Credit: NASA)

Close your eyes and imagine the universe not as it is today, filled with galaxies, stars, and planets, but as it was 13.8 billion years ago—a single point of infinite density, infinite temperature, and infinite possibility. This is the Big Bang, the moment when time and space themselves burst into existence. It's not an explosion in the conventional sense, but the rapid expansion of the very fabric of reality. In less than a second, the universe grew from smaller than an atom to the size of a grapefruit, and it hasn't stopped expanding since.

But how do we know this happened? How can we be sure that the universe had a beginning, let alone one so dramatic? The answers lie in the stars, in the light that has traveled billions of years to reach us, and in the faint echoes of that primordial explosion that still resonate through the cosmos.

The Redshift: A Universe on the Move

In the 1920s, a young astronomer named Edwin Hubble peered through the lens of the most powerful telescope of his time, the 100-inch Hooker Telescope at Mount Wilson Observatory, and made a discovery that would forever alter our understanding of the cosmos. What he saw was not just a static sea of stars but a dynamic, ever-changing universe—one that was alive with motion, expansion, and mystery. His observations would lead to one of the most profound revelations in the history of science: the universe is expanding.

Hubble's breakthrough began with something seemingly mundane: the light from distant galaxies. When he analyzed this light, he noticed something peculiar. The light from these galaxies was shifted toward the red end of the spectrum. This phenomenon, known as redshift, occurs when an object is moving away from an observer, stretching the wavelength of its light and making it appear redder. It's the same effect that causes the sound of a passing ambulance siren to drop in pitch as it moves away from you—a cosmic Doppler effect.

But Hubble didn't stop there. He meticulously measured the redshifts of dozens of galaxies and made an even more astonishing discovery: the farther away a galaxy was, the greater its redshift. In other words, the more distant a galaxy, the faster it was receding from us. This relationship, now enshrined as Hubble's Law, revealed something extraordinary: the universe itself is expanding.

To grasp the enormity of this discovery, imagine a loaf of raisin bread baking in the oven. As the dough rises, the raisins move apart from one another. If you were a raisin, you'd see all the other raisins moving away from you, with the ones farther away moving faster. This is precisely what Hubble observed in the cosmos. The galaxies are like the raisins, and the expanding universe is the rising dough. No matter where you are in the loaf—or the universe—you'd see the same thing: everything moving away from you, with more distant objects receding faster.

This analogy isn't just a helpful visualization; it's a profound insight into the nature of the universe. The expansion isn't happening *into* anything—it's the very fabric of space itself that's stretching. And if the universe is expanding now, it must have been smaller in the past. Rewind the tape far enough, and you reach a point where everything was compressed into an infinitely dense, infinitely hot state—the *Big Bang*.

Hubble's discovery was revolutionary because it provided the first observational evidence that the universe had a beginning. Before Hubble, the prevailing view was that the universe was static and eternal. But his observations shattered that notion, revealing a cosmos that was dynamic, evolving, and finite in age. This revelation was so profound that even Albert Einstein, who had initially resisted the idea of an expanding universe, called his earlier skepticism his "greatest blunder."

Hubble's work didn't just change our understanding of the universe; it changed the way we see ourselves within it. We are not living in a static, unchanging cosmos but in a universe that is alive with motion, growth, and transformation. And if the universe is expanding, it raises an even bigger question: where is it going? What is its ultimate fate?

Hubble's Law didn't just tell us that the universe is expanding; it gave us a way to measure that expansion. By determining the redshift of a galaxy, astronomers can calculate how fast it's moving away from us. And by combining this information with other measurements, such as the brightness of supernovae, they can map the expansion of the universe over time.

But Hubble's discovery also opened the door to even deeper mysteries. If the universe is expanding, what's driving that expansion? And will it continue forever, or will it eventually slow down, stop, or even reverse? These questions would lead to the discovery of dark energy, a mysterious force that's accelerating the expansion of the universe, and to the development of theories about the ultimate fate of the cosmos—the Big Crunch, the Big Freeze, and the Big Rip.

Hubble's discovery of the expanding universe is more than just a scientific milestone; it's a reminder of our place in the cosmos. We are tiny beings on a small planet, orbiting an ordinary star in a vast galaxy, which is itself just one of billions in an ever-expanding universe. And yet, with our curiosity and ingenuity, we've been able to uncover some of the deepest secrets of the cosmos.

As we stand on the shoulders of giants like Hubble, we continue to explore the universe, pushing the boundaries of our knowledge and imagination. The story of the expanding universe is not just a story of galaxies and redshifts; it's a story of discovery, wonder, and the relentless human quest to understand the cosmos.

The Echo of Creation

The story of the Big Bang doesn't end with Edwin Hubble's groundbreaking discovery of the expanding universe. While Hubble's observations provided the first clues that the universe had a beginning, it wasn't until the mid-20th century that scientists found what many consider the smoking gun: the *Cosmic Microwave Background (CMB)*, the faint afterglow of the Big Bang itself. This discovery, made almost by accident, would become one of the most profound confirmations of the Big Bang theory and a cornerstone of modern cosmology.

A Serendipitous Discovery

In 1965, two radio astronomers, *Arno Penzias* and *Robert Wilson*, were working at Bell Labs in New Jersey, using a large horn-shaped antenna to study radio waves from space. Their goal was to measure signals from celestial objects, but they kept encountering a persistent problem: a faint, uniform noise that seemed to come from every direction in the sky. At first, they assumed it was interference—perhaps from nearby radio towers, atmospheric disturbances, or even pigeons nesting in their antenna. They meticulously cleaned the antenna, shooed away the pigeons, and checked every possible source of interference, but the noise refused to go away.

Frustrated and puzzled, Penzias and Wilson reached out to a group of physicists at Princeton University, led by *Robert Dicke*, who had been theorizing about the possibility of detecting leftover radiation from the Big Bang. When the two teams compared notes, the truth became clear: Penzias and Wilson had stumbled upon something extraordinary. The noise they had been trying to eliminate was not interference—it was the *Cosmic Microwave Background*, the faint echo of the universe's birth.

The Oldest Light in the Universe

The CMB is the oldest light we can observe, a snapshot of the universe when it was just *380,000 years old*. At that time, the universe was a hot, dense soup of particles and energy, too hot for atoms to form. Electrons and protons moved freely, scattering light and making the universe opaque. But as the universe expanded and cooled, it reached a critical temperature—about *3,000 degrees Celsius*—allowing electrons and protons to combine into neutral hydrogen atoms. This moment, known as recombination, marked the first time light could travel freely through the universe.

The light released during recombination has been traveling through space ever since, stretched by the expansion of the universe from visible wavelengths into the microwave part of the spectrum. Today, this ancient light fills the entire cosmos, a faint glow that permeates every corner of the universe. It is the CMB, a cosmic fossil that carries with it the secrets of the universe's infancy.

A Cosmic Fingerprint

The CMB is more than just a relic of the past; it is a treasure trove of information about the early universe. Its temperature is remarkably uniform, hovering at just *2.725 degrees above absolute zero*, but it is not perfectly smooth. Tiny fluctuations in temperature—on the order of one part in 100,000—reveal the seeds of structure in the early universe. These fluctuations are the imprints of density variations, regions where matter was slightly more or less dense than average. Over billions of years, these tiny ripples would grow under the influence of gravity, eventually forming the galaxies, stars, and planets we see today.

The uniformity of the CMB supports the idea of cosmic inflation, a theory proposing that the universe underwent an exponential expansion in the first fraction of a second after the Big Bang. Inflation would have stretched the universe to an enormous size, smoothing out irregularities and setting the stage for the formation of large-scale structures. Without inflation, the universe would not be as uniform as we observe it to be.

The discovery of the CMB was a triumph for the Big Bang theory, but it also raised new questions. Why is the CMB so uniform? What caused the tiny fluctuations in temperature? And what does the CMB tell us about the composition and evolution of the universe?

To answer these questions, scientists have launched a series of increasingly sophisticated missions to study the CMB in detail. The Cosmic Background Explorer (COBE) satellite, launched in 1989, provided the first precise measurements of the CMB's temperature and confirmed the existence of the tiny fluctuations predicted by theory. Later missions, such as the *Wilkinson Microwave Anisotropy Probe (WMAP)* and the *Planck satellite*, mapped the CMB with even greater precision, revealing a wealth of information about the universe's age, composition, and evolution.

These observations have confirmed that the universe is 13.8 billion years old and composed of about *5% ordinary matter*, *27% dark matter*, and *68% dark energy*. They have also provided evidence for cosmic inflation and shed light on the conditions that gave rise to the first galaxies and stars.

The discovery of the CMB is one of the greatest achievements in the history of science, a testament to the power of curiosity, collaboration, and perseverance. It has transformed our understanding of the universe, providing a direct link to its origins and a window into its deepest mysteries. But the CMB is more than just a scientific milestone; it is a reminder of our connection to the cosmos.

When we look at the CMB, we are seeing the universe as it was just 380,000 years after the Big Bang, a time when the first atoms were forming and light was beginning to shine. We are witnessing the dawn of creation, the moment when the universe transitioned from a hot, opaque plasma to a transparent, expanding cosmos. And we are seeing the seeds of everything that would come after—the galaxies, stars, planets, and, ultimately, us.

Dark Energy: The Cosmic Accelerator

Just when we thought we had a handle on the universe's expansion, cosmologists were hit with a curveball that would upend everything we thought we knew. In the late 1990s, two independent teams of astronomers made a discovery so shocking that it would earn them the *Nobel Prize in Physics* in 2011. By studying distant supernovae—exploding stars that serve as cosmic mile markers—they found that the universe wasn't just expanding; it was accelerating. Something mysterious was pushing galaxies apart faster and faster, as if the cosmos had stepped on the gas pedal. This mysterious force was dubbed *dark energy*, and it remains one of the greatest enigmas in modern science.

The Supernova Surprise

The discovery of dark energy began with a quest to measure the universe's expansion rate. Astronomers had long known that the universe was expanding, thanks to Edwin Hubble's groundbreaking work in the 1920s.

But the question remained: was this expansion slowing down over time, as gravity pulled galaxies together, or was it speeding up? To find out, scientists turned to *Type Ia supernovae*, a special class of exploding stars that shine with a consistent brightness. Because their intrinsic brightness is known, these supernovae can be used as "standard candles" to measure cosmic distances.

When the two teams—the *Supernova Cosmology Project* and the *High-Z Supernova Search Team*—compared the distances of these supernovae to their redshifts, they expected to find that the expansion of the universe was slowing down. Instead, they found the opposite: the supernovae were fainter than expected, meaning they were farther away than predicted. The universe wasn't just expanding; it was accelerating.

This was a revelation that no one had anticipated. It was as if you threw a ball into the air, expecting it to fall back down, only to watch it shoot upward faster and faster. Something was counteracting gravity, pushing the universe apart with increasing speed. That something was dark energy.

The Mystery of Dark Energy

Dark energy is the name we give to this mysterious force, but what exactly is it? The truth is, we don't know. Dark energy makes up about *68% of the universe,* yet it is completely invisible and undetectable by conventional means. It doesn't emit light, absorb light, or interact with matter in any way we can observe directly. All we know is that it's there, and it's powerful enough to dominate the universe's evolution.

The leading theory is that dark energy is a property of space itself, often referred to as the *cosmological constant* (represented by the Greek letter Λ). This idea dates back to Albert Einstein, who introduced the concept in 1917 to explain a static universe. When Hubble discovered the universe was expanding, Einstein abandoned the cosmological constant, calling it his "greatest blunder." But now, it seems, Einstein may have been right after all—just for the wrong reasons.

If dark energy is indeed the cosmological constant, it means that empty space is not truly empty. Instead, it has an inherent energy that causes it to expand. This energy is constant over time and uniform throughout space, acting as a kind of anti-gravity that pushes galaxies apart. But this is just one possibility. Dark energy could also be something more exotic, like a dynamic field that changes over time, known as *quintessence*, or even a sign that our understanding of gravity is incomplete.

Dark energy doesn't just explain the universe's current acceleration; it holds the key to its ultimate fate. Depending on the nature of dark energy, the universe could end in one of several dramatic scenarios:

- *The Big Freeze:* If dark energy remains constant, the universe will continue to expand at an accelerating rate. Over time, galaxies will move farther apart, stars will burn out, and the universe will grow colder and darker, eventually reaching a state of maximum entropy known as the *heat death*.
- *The Big Rip:* If dark energy grows stronger over time, it could eventually overcome all other forces, tearing apart galaxies, stars, planets, and even atoms in a catastrophic event known as the Big Rip.
- *The Big Crunch:* If dark energy weakens or reverses, gravity could eventually take over, causing the universe to collapse back in on itself in a fiery Big Crunch, potentially leading to a new Big Bang and a cyclic universe.

Understanding dark energy is one of the greatest challenges in modern cosmology. To unravel its mysteries, scientists are using a variety of tools and techniques, from ground-based telescopes to space observatories. Missions like the *Dark Energy Survey (DES)*, the *Euclid Space Telescope*, and the *Nancy Grace Roman Space Telescope* are designed to map the distribution of galaxies and measure the universe's expansion with unprecedented precision.

These efforts aim to answer some of the most pressing questions about dark energy: Is it truly constant, or does it change over time? How does it interact with dark matter, the other invisible component of the universe? And what does it tell us about the fundamental laws of physics?

The Stage is Set

Dark energy is more than just a scientific mystery; it is a reminder of how much we still have to learn about the universe. It challenges our understanding of space, time, and gravity, and it forces us to confront the limits of our knowledge. But it also inspires us to keep exploring, to keep asking questions, and to keep pushing the boundaries of what we think is possible.

As we stand on the precipice of discovery, staring into the vast unknown, we are reminded that the universe is full of surprises. Dark energy is one of its greatest secrets, a cosmic puzzle that defies explanation. But it is also a beacon, guiding us toward a deeper understanding of the cosmos and our place within it.

The Big Bang set the universe in motion, and dark energy is shaping its future. But what does this mean for the ultimate fate of the universe? Will the expansion continue forever, leading to a cold, dark end? Or will the universe collapse back in on itself in a fiery Big Crunch? Or perhaps, something even more dramatic awaits us—a Big Rip that tears the very fabric of reality apart.

In the chapters that follow, we'll explore these possibilities, diving into the science, the theories, and the profound implications of each scenario. The journey begins here, at the dawn of the universe, where the seeds of its destiny were sown.

Chapter 2: The Big Crunch – A Cyclic Universe

Imagine the universe as a pendulum, swinging back and forth between creation and destruction. On one side of the swing lies the Big Bang, the moment of fiery birth. On the other side lies the Big Crunch, the moment of cosmic collapse. This is the vision of a cyclic universe, one that expands, contracts, and begins anew in an endless dance of creation and destruction. It's a vision that has captivated scientists and philosophers alike, offering a dramatic alternative to the cold, dark fate of the Big Freeze.

But is the Big Crunch a real possibility, or is it just a cosmic fantasy? To answer this question, we must delve into the forces that shape the universe, from the relentless pull of gravity to the mysterious push of dark energy.

The Gravity of the Situation

Gravity is the great architect of the cosmos. It is the force that sculpts galaxies, binds stars into clusters, and keeps planets in their orbits. It is the force that pulls us to the ground and holds the universe together. But gravity is also the force that could, one day, bring the universe's expansion to a screeching halt and reverse it, leading to a dramatic and fiery end known as the Big Crunch. To understand how this could happen, we must first understand the nature of gravity and its role in the cosmic drama.

Gravity is one of the four fundamental forces of nature, alongside electromagnetism, the strong nuclear force, and the weak nuclear force. But unlike the other forces, which operate on microscopic scales or within specific contexts, gravity is universal. It acts on everything that has mass, from the tiniest particle to the largest galaxy. It is the force that shapes the large-scale structure of the universe, pulling matter together to form stars, galaxies, and clusters of galaxies.

The story of gravity begins with Isaac Newton, who first described it as a force that attracts objects to one another. His law of universal gravitation explained how the same force that causes an apple to fall from a tree also keeps the Moon in orbit around the Earth. But it wasn't until Albert Einstein's theory of general relativity that we truly understood the profound nature of gravity. Einstein showed that gravity is not just a force; it is the curvature of spacetime itself. Massive objects like stars and planets warp the fabric of spacetime, and this curvature dictates how objects move.

Gravity vs. Expansion

In the early days of cosmology, scientists believed that gravity would eventually slow down the universe's expansion. After all, gravity is an attractive force, and over time, it should pull galaxies back together. The universe's expansion, driven by the momentum of the Big Bang, was thought to be like a ball thrown into the air. Eventually, gravity would overcome the ball's upward motion, bringing it back down to Earth. Similarly, gravity was expected to overcome the universe's outward expansion, causing it to collapse in on itself.

This idea was rooted in the concept of the critical density of the universe. If the universe's density—the amount of matter and energy it contains—was above a certain critical value, gravity would eventually win out, and the universe would collapse. If the density was below this value, the universe would continue to expand forever. For much of the 20th century, scientists believed that the universe was teetering on the edge of this critical density, perfectly balanced between expansion and collapse.

The possibility of a Big Crunch depends on the amount of matter in the universe. Matter, whether it's the visible stars and galaxies or the invisible dark matter, exerts a gravitational pull that counteracts the universe's expansion. The more matter there is, the stronger the gravitational pull, and the greater the chance of a collapse.

In the 20th century, astronomers worked tirelessly to measure the universe's density. They counted galaxies, mapped their distribution, and studied the cosmic microwave background (CMB) to determine how much matter the universe contains. For a time, it seemed that the universe might have just enough matter to bring about a Big Crunch. But then came the discovery of dark energy, and everything changed.

The Dark Energy Factor

Dark energy is the wildcard in the cosmic deck. It is the mysterious force that is driving the universe's accelerated expansion, counteracting the pull of gravity. If dark energy remains constant or grows stronger over time, it will continue to push galaxies apart, making a Big Crunch impossible. But if dark energy weakens or reverses, gravity could regain the upper hand, leading to a cosmic collapse.

This uncertainty is what makes the Big Crunch such a tantalizing possibility. It's a reminder that the universe's fate is not yet written, that the forces shaping it are still not fully understood. And it raises profound questions about the nature of dark energy and its role in the cosmos.

If gravity does eventually overcome the universe's expansion, the result would be a Big Crunch—a fiery end that mirrors the Big Bang in its intensity. As galaxies are pulled closer together, the universe would grow hotter and denser. Stars would collide, planets would be torn apart, and matter would be compressed into an increasingly smaller space. Eventually, the entire universe would collapse into a singularity, a point of infinite density and temperature.

But the Big Crunch isn't just an end; it's also a beginning. In some theories, the collapse of the universe could trigger a new Big Bang, leading to a new cycle of expansion and contraction. This idea, known as the cyclic universe or oscillating universe model, suggests that the universe goes through endless cycles of birth, death, and rebirth.

The Critical Density

The fate of the universe hinges on a delicate balance—a cosmic tightrope walk between expansion and collapse. This balance is determined by a single, crucial factor: the density of the universe. If the universe's density is above a certain critical value, gravity will eventually overcome the outward momentum of the Big Bang, pulling everything back together in a fiery Big Crunch. If the density is below this value, the universe will continue to expand forever, fading into cold, dark emptiness. This critical density is the dividing line between two radically different cosmic destinies, and understanding it is key to unraveling the ultimate fate of the cosmos.

The concept of critical density is rooted in the interplay between two fundamental forces: the outward push of the Big Bang and the inward pull of gravity. The Big Bang set the universe in motion, sending galaxies flying apart like shrapnel from an explosion. But gravity, the great cosmic glue, works to pull everything back together. The question is: which force will win out in the end?

The critical density is the tipping point in this cosmic tug-of-war. It is the amount of matter and energy required to exactly balance the universe's expansion. If the universe's density is equal to the critical density, the expansion will slow down over time but never quite stop, resulting in a *flat universe.* If the density is higher, gravity will eventually reverse the expansion, leading to a Big Crunch. If the density is lower, the universe will expand forever, growing colder and darker as it does.

The Search for Omega

To determine whether the universe's density is above, below, or equal to the critical density, scientists use a parameter called *Omega (Ω)*. Omega is the ratio of the universe's actual density to the critical density. If Omega is greater than 1, the universe is *closed* and will eventually collapse. If Omega is less than 1, the universe is *open* and will expand forever. If Omega is exactly 1, the universe is *flat*, balanced on the knife's edge between expansion and collapse.

For much of the 20th century, scientists believed that Omega was very close to 1, meaning the universe was perfectly balanced. This idea, known as the flat universe scenario, was supported by early observations of the cosmic microwave background (CMB) and the distribution of galaxies. The CMB, the afterglow of the Big Bang, showed a remarkably uniform temperature, suggesting that the universe was finely tuned to avoid either a rapid collapse or a runaway expansion.

The Flat Universe Puzzle

The flat universe scenario was not just a scientific hypothesis; it was a philosophical puzzle. Why should the universe be so perfectly balanced? Was it a coincidence, or was there some deeper reason? Some scientists speculated that the universe's flatness was the result of a process called cosmic inflation, a rapid expansion that occurred in the first fraction of a second after the Big Bang. Inflation would have stretched the universe to an enormous size, smoothing out irregularities and setting the stage for the formation of galaxies and large-scale structures.

But while inflation explained the universe's flatness, it also raised new questions. If the universe was so finely tuned, where was all the matter needed to achieve critical density? Observations of visible matter—stars, galaxies, and gas—accounted for only a small fraction of the required density. The rest, it seemed, was missing.

In the late 1990s, two independent teams of astronomers made a discovery that would change everything. By studying distant Type Ia supernovae, they found that the universe's expansion was not slowing down, as expected, but accelerating. This acceleration could only be explained by the presence of a mysterious force pushing galaxies apart—a force that came to be known as dark energy.

Dark energy makes up about 68% of the universe's density, more than enough to account for the missing mass needed to achieve critical density. But dark energy is not matter; it is a property of space itself, a kind of anti-gravity that drives the universe's accelerated expansion. Its discovery solved the flatness puzzle but also introduced a new mystery: what is dark energy, and how does it work?

The discovery of dark energy has profound implications for the fate of the universe. If dark energy remains constant or grows stronger over time, it will continue to push galaxies apart, making a Big Crunch impossible. But if dark energy weakens or reverses, gravity could regain the upper hand, leading to a cosmic collapse.

This uncertainty is what makes the critical density such a fascinating concept. It is a reminder that the universe's fate is not yet written, that the forces shaping it are still not fully understood. And it raises profound questions about the nature of dark energy, the role of gravity, and the ultimate destiny of the cosmos.

The Dark Energy Wildcard

Just when cosmologists thought they had a handle on the universe's fate, dark energy emerged as the ultimate wildcard—a mysterious force that defies intuition and upends our understanding of the cosmos. Instead of slowing down, as gravity's relentless pull would suggest, the universe's expansion is speeding up, driven by this enigmatic energy. Dark energy doesn't just challenge the idea of a Big Crunch; it rewrites the rules of the cosmic game, leaving scientists grappling with questions that stretch the limits of physics and imagination.

The simplest explanation for dark energy is the *cosmological constant*, a term Albert Einstein introduced in 1917 to describe a static universe. When Hubble discovered the universe was expanding, Einstein abandoned the idea, calling it his "greatest blunder." But now, it seems, Einstein may have been right after all—just for the wrong reasons. The cosmological constant represents a constant energy density filling space, driving the universe's accelerated expansion.

If dark energy is indeed the cosmological constant, it means that empty space is not truly empty. Instead, it has an inherent energy that causes it to expand. This energy is uniform throughout space and constant over time, acting as a steady pressure that counteracts gravity. But this is just one possibility. Dark energy could also be something more dynamic, a field that changes over time, known as *quintessence*, or even a sign that our understanding of gravity is incomplete.

The Big Crunch's Last Stand

Dark energy's discovery threw a wrench into the Big Crunch scenario. If dark energy remains constant or grows stronger over time, it will continue to push galaxies apart, making a cosmic collapse impossible. The universe would expand forever, growing colder and darker as it does—a fate known as the Big Freeze. But if dark energy weakens or reverses, gravity could regain the upper hand, pulling the universe back together in a fiery Big Crunch.

This uncertainty is what makes the Big Crunch such a tantalizing possibility. It's a reminder that the universe's fate is not yet written, that the forces shaping it are still not fully understood. And it raises profound questions: Is dark energy truly constant, or is it evolving? Could it grow stronger, eventually tearing the universe apart in a Big Rip? Or could it fade away, allowing gravity to take control once more?

The Phantom Menace

One of the most dramatic possibilities is that dark energy could grow stronger over time, leading to a Big Rip. In this scenario, dark energy would eventually overcome all other forces, tearing apart galaxies, stars, planets, and even atoms. The universe would end not with a whimper, but with a catastrophic rending of spacetime itself. This idea, while speculative, highlights the strange and unpredictable nature of dark energy.

On the other hand, if dark energy weakens, it could allow gravity to reverse the universe's expansion, leading to a Big Crunch. In this scenario, galaxies would slow down, stop, and then begin to move toward each other. The universe would grow hotter and denser, eventually collapsing into a singularity—a point of infinite density and temperature. This fiery end could also mark a new beginning, triggering another Big Bang and restarting the cosmic cycle.

The Cyclic Universe

The Big Crunch isn't just an end; it's also a beginning. Imagine a universe that doesn't simply fade into cold darkness or tear itself apart in a violent rip, but instead collapses in on itself, only to explode outward once again in a fiery rebirth. This is the vision of the cyclic universe, a cosmos that goes through endless cycles of expansion and contraction, birth and death, creation and destruction. It's a vision of eternity, a universe that never truly ends but is constantly reborn.

In the cyclic universe model, each cycle begins with a Big Bang and ends with a Big Crunch. The universe expands, cools, and evolves, forming galaxies, stars, and planets. But eventually, gravity takes over, pulling everything back together in a fiery collapse. The universe becomes a hot, dense singularity, a point of infinite density and temperature. And then, in a cosmic reset, it explodes outward once again, starting the cycle anew.

Each cycle could last billions or even trillions of years, with the universe evolving in different ways each time. Perhaps the laws of physics change slightly from one cycle to the next, leading to new forms of matter, energy, and even life. Or perhaps the universe remains fundamentally the same, repeating its history over and over in an endless loop. The possibilities are as vast as the cosmos itself.

The Appeal of Eternity

The idea of a cyclic universe has a certain poetic appeal. It suggests that the universe is not a one-time event but an eternal process, a cosmic dance of creation and destruction. It's a vision that blurs the line between beginnings and endings, suggesting that every end is also a beginning. And it raises profound questions about the nature of time, space, and existence.

But the cyclic universe is more than just a philosophical idea; it's a scientific hypothesis that has been explored by some of the greatest minds in physics. Albert Einstein himself toyed with the idea of a cyclic universe, as did Richard Tolman and Paul Steinhardt, who developed modern versions of the theory. These theories suggest that the universe could undergo endless cycles of expansion and contraction, each one leaving behind subtle traces of the previous cycle.

While the cyclic universe is an intriguing idea, it's not without its challenges. One of the biggest is the second law of thermodynamics, which states that entropy—a measure of disorder—always increases over time. If the universe goes through endless cycles, each one would add to the entropy of the cosmos, leading to a universe that becomes more and more disordered with each cycle. This raises the question: how could the universe reset itself without violating the laws of physics?

Another challenge is the nature of the Big Crunch itself. If the universe collapses into a singularity, what happens to the information and structure of the previous cycle? Does it get erased, or is it somehow preserved? These are questions that push the boundaries of our understanding of physics and cosmology.

The cyclic universe is a reminder that the cosmos is full of surprises, that the story of the universe is far from over. It's a vision that challenges our understanding of time, space, and existence, and it raises profound questions about the nature of reality.

The Evidence

So, is there any evidence for the Big Crunch? The answer is complicated. On the one hand, the discovery of dark energy and the accelerating expansion of the universe make a Big Crunch seem unlikely. If dark energy remains constant or grows stronger over time, it will continue to push galaxies apart, making a collapse impossible. But on the other hand, our understanding of dark energy is still incomplete, and it's possible that its behavior could change in the future.

Dark energy is the great unknown in cosmology. We know it makes up about 68% of the universe's energy density, and we know it's driving the accelerated expansion of the universe. But we don't know what it is, how it works, or whether it will remain constant over time. Some theories suggest that dark energy could decay over time, weakening its repulsive force and allowing gravity to take over. Others propose that dark energy could interact with dark matter, the invisible substance that makes up about 27% of the universe, in ways that could lead to a collapse.

Another possibility is that quantum gravity—a theory that unites general relativity with quantum mechanics—could play a role in the universe's fate. At the incredibly high densities and temperatures of a Big Crunch, quantum effects would dominate, potentially preventing the universe from collapsing into a singularity. Instead, the universe could bounce back, leading to a new Big Bang and a new cycle of expansion.

This idea, known as the Big Bounce, suggests that the universe could undergo endless cycles of expansion and contraction, each one leaving behind subtle traces of the previous cycle. While the Big Bounce is still a speculative idea, it offers a tantalizing possibility: that the universe is not a one-time event but an eternal process, a cosmic cycle of birth, death, and rebirth.

The Search for Clues

So, how can we find evidence for the Big Crunch or the cyclic universe? One way is to study the cosmic microwave background (CMB), the afterglow of the Big Bang. The CMB contains subtle clues about the early universe, including the density of matter and the rate of expansion. By studying these clues, scientists can learn more about the universe's fate.

Another way is to study dark energy and its behavior over time. If dark energy is changing, it could leave behind subtle traces in the distribution of galaxies and the large-scale structure of the universe. By mapping these structures, scientists can learn more about the forces shaping the cosmos.

The search for evidence of the Big Crunch is a reminder of how much we still have to learn about the universe. It's a journey that takes us to the edge of reality, to the limits of our understanding, and to the heart of the cosmic mystery. Whether the universe ends in fire, ice, or something else entirely, the quest to uncover the truth is one of the greatest adventures in human history.

The Implications

The Big Crunch is more than just a scientific hypothesis; it's a profound idea that challenges our understanding of time, space, and existence. It's a concept that forces us to confront the very nature of reality, to grapple with questions that stretch the limits of human imagination. If the universe is cyclic, then everything we know—every galaxy, every star, every planet, every life—is part of an endless cycle of creation and destruction. It's a vision that blurs the line between beginnings and endings, suggesting that every end is also a beginning.

A Universe in Flux

Imagine a universe that doesn't simply fade into cold darkness or tear itself apart in a violent rip, but instead collapses in on itself, only to explode outward once again in a fiery rebirth. This is the vision of the cyclic universe, a cosmos that goes through endless cycles of expansion and contraction, birth and death, creation and destruction. It's a vision of eternity, a universe that never truly ends but is constantly reborn.

In this model, each cycle could last billions or even trillions of years, with the universe evolving in different ways each time. Perhaps the laws of physics change slightly from one cycle to the next, leading to new forms of matter, energy, and even life. Or perhaps the universe remains fundamentally the same, repeating its history over and over in an endless loop. The possibilities are as vast as the cosmos itself.

The Nature of Time

The Nature of Time The Big Crunch raises profound questions about the nature of time. If the universe collapses into a singularity, does time itself come to an end? Or does it simply reset, starting anew with the next Big Bang? These are questions that push the boundaries of our understanding of physics and philosophy.

In a cyclic universe, time could be seen as a series of loops, each one beginning with a Big Bang and ending with a Big Crunch. But what happens to the information and structure of the universe during the collapse? Does it get erased, or is it somehow preserved? These are questions that challenge our understanding of entropy, information theory, and the very fabric of reality.The Big Crunch raises profound questions about the nature of time. If the universe collapses into a singularity, does time itself come to an end?

Or does it simply reset, starting anew with the next Big Bang? These are questions that push the boundaries of our understanding of physics and philosophy.

In a cyclic universe, time could be seen as a series of loops, each one beginning with a Big Bang and ending with a Big Crunch. But what happens to the information and structure of the universe during the collapse? Does it get erased, or is it somehow preserved? These are questions that challenge our understanding of entropy, information theory, and the very fabric of reality.

The Fate of Information

One of the most intriguing questions raised by the Big Crunch is the fate of information. In our current understanding of physics, information is never truly lost; it is simply transformed. But in a Big Crunch, the universe collapses into a singularity, a point of infinite density and temperature. What happens to the information contained within the universe? Does it get compressed into the singularity, or is it somehow preserved for the next cycle?

This question is not just a theoretical curiosity; it has profound implications for our understanding of the universe. If information is preserved, it could mean that each cycle of the universe leaves behind subtle traces of the previous cycle. These traces could be encoded in the cosmic microwave background, the distribution of galaxies, or even the fundamental constants of nature. By studying these traces, we could learn more about the history of the universe and the nature of reality itself.

Existential Questions

The Big Crunch also raises existential questions about our place in the cosmos. If the universe is cyclic, then everything we know—every galaxy, every star, every planet, every life—is part of an endless cycle of creation and destruction.

It's a vision that challenges our understanding of existence, forcing us to confront the impermanence of everything we hold dear.

But it's also a vision that offers a kind of cosmic comfort. If the universe is cyclic, then every end is also a beginning. The collapse of the universe is not just a fiery end; it's also the birth of a new cosmos, a new cycle of expansion and contraction. In this vision, the universe is not a one-time event but an eternal process, a cosmic dance of creation and destruction.

A Cosmic Drama

The story of the Big Crunch is a story of cosmic drama, of forces locked in an eternal struggle between expansion and contraction, creation and destruction. It's a story that reminds us of the fragility and resilience of the universe, of its capacity for both growth and collapse.

As we continue to explore the mysteries of the cosmos, the Big Crunch remains a tantalizing possibility, a reminder that the universe's fate is still uncertain. Whether it ends in fire, ice, or something else entirely, the journey to uncover the truth is one of the greatest adventures in human history.

And so, as we stand on the edge of the unknown, we are reminded that the universe is full of surprises. The Big Crunch may be just one of many possible endings, but it is a reminder that the story of the cosmos is far from over. The pendulum continues to swing, and the universe continues to unfold, revealing its secrets one discovery at a time.

Chapter 3: The Big Freeze – Heat Death of the Universe

Imagine a universe that doesn't end in fire or collapse, but in ice—a vast, cold, and empty expanse where stars have burned out, galaxies have drifted apart, and all that remains is a faint, fading glow of what once was. This is the *Big Freeze*, also known as the heat death of the universe, a scenario in which the cosmos continues to expand forever, growing colder and darker until it reaches a state of maximum entropy. It's a vision of the universe's end that is as haunting as it is inevitable, a reminder that even the most vibrant and dynamic systems must eventually succumb to the relentless march of time.

But what exactly is the Big Freeze, and why is it considered one of the most likely fates for our universe? To understand this, we must delve into the science of thermodynamics, the nature of entropy, and the mysterious force driving the universe's accelerated expansion: *dark energy*.

The Second Law of Thermodynamics

The Big Freeze is rooted in one of the most fundamental laws of physics: the second law of thermodynamics. This law states that in any closed system, *entropy*—a measure of disorder—always increases over time. In simpler terms, things tend to move from order to disorder, from energy concentrated in one place to energy spread out evenly. It's a law that governs everything from the cooling of a cup of coffee to the fate of the entire universe.

To understand entropy, imagine a drop of ink falling into a glass of water. At first, the ink is concentrated in one spot, a small island of color in a sea of clear liquid. But over time, the ink spreads out, diffusing through the water until it is evenly distributed. This is entropy at work: the system moves from a state of order (the ink concentrated in one place) to a state of disorder (the ink spread evenly throughout the water).

Now, imagine this process on a cosmic scale. The universe began in a state of incredible order—a hot, dense, and nearly uniform soup of particles and energy. But as the universe expanded and cooled, it moved toward greater disorder. Stars formed, burned out, and scattered their elements across the cosmos. Galaxies collided and merged, spreading matter even further. And all the while, entropy increased, driving the universe toward a state of maximum disorder.

In the context of the universe, the second law of thermodynamics means that over time, all the energy and matter in the cosmos will become increasingly dispersed. Stars will burn out, galaxies will drift apart, and the universe will grow colder and darker. Eventually, it will reach a state of maximum entropy, where no usable energy remains to power any processes—cosmic, chemical, or biological. This is the *heat death of the universe*, the ultimate triumph of entropy over order.

The Inevitability of Entropy

The inevitability of entropy is not just a theoretical concept; it is supported by a wealth of observational data. For example, studies of the cosmic microwave background (CMB), the afterglow of the Big Bang, show that the early universe was incredibly uniform, with only tiny fluctuations in temperature and density. These fluctuations were the seeds of structure in the universe, the regions where matter would eventually clump together to form galaxies and stars.

But as the universe has expanded, these structures have become increasingly dispersed. Observations of distant galaxies show that they are moving away from us at ever-increasing speeds, driven by the mysterious force of dark energy. As the universe continues to expand, the space between galaxies will grow, and the density of matter and energy will decrease. Eventually, the universe will reach a state where no new stars can form, and existing stars will burn out, leaving behind only cold, dark remnants.

Research data from missions like the Wilkinson Microwave Anisotropy Probe (WMAP) and the Planck satellite have provided precise measurements of the universe's expansion rate and the distribution of matter and energy. These measurements confirm that dark energy is driving the accelerated expansion of the universe, making the Big Freeze one of the most likely scenarios for the universe's ultimate fate.

A Universe in Isolation

As the universe expands, driven by the relentless force of dark energy, galaxies will become increasingly isolated. The space between them will grow so vast that light from one galaxy will never reach another. Each galaxy will become its own island universe, cut off from the rest of the cosmos. Over time, even these islands will fade as stars die and black holes evaporate through a process known as *Hawking radiation*. The universe will become a cold, dark, and lifeless expanse, a shadow of its former vibrant self.

The Expanding Cosmos: A Growing Divide

The expansion of the universe is not just a theoretical concept; it is a reality supported by decades of observational data. Edwin Hubble's discovery of redshift in the 1920s first revealed that galaxies are moving away from us, and subsequent studies have shown that this expansion is accelerating. According to data from the Hubble Space Telescope and other observatories, the universe's expansion rate (the *Hubble Constant*) is approximately *73 kilometers per second per megaparsec*. This means that for every 3.26 million light-years of distance, galaxies are receding from us at an additional 73 kilometers per second.

As dark energy continues to drive this accelerated expansion, the distances between galaxies will grow exponentially. Today, the nearest major galaxy to the Milky Way, Andromeda, is about 2.5 million light-years away. In the distant future, even Andromeda—currently on a collision course with our galaxy—will be pulled away by the expansion of space, leaving the Milky Way as an isolated island in a vast cosmic ocean.

The Fate of Galaxies: Islands in the Void

In the far future, galaxies will no longer be able to interact with one another. The space between them will stretch to such an extent that light from one galaxy will never reach another. This isolation will have profound consequences for the evolution of galaxies. Without the gravitational influence of neighboring galaxies, star formation will cease as the raw materials—gas and dust—are spread too thin. The universe will become a collection of isolated islands, each one fading into darkness.

Even within these isolated galaxies, the end is inevitable. Stars will exhaust their nuclear fuel, burning out one by one. Massive stars will end their lives in spectacular supernovae, leaving behind *neutron stars* or *black holes.* Smaller stars, like our Sun, will swell into red giants before shedding their outer layers and collapsing into *white dwarfs,* which will slowly cool and fade over trillions of years.

In the final stages of the Big Freeze, the universe will be dominated by black holes. These remnants of dead stars and galaxies will be the last bastions of structure in an otherwise empty cosmos. But even black holes are not eternal. Through a process predicted by Stephen Hawking in 1974, known as *Hawking radiation,* black holes will slowly lose mass and energy, evaporating over unimaginable timescales.

For a stellar-mass black hole, this process takes about *10^67 years.* For supermassive black holes, it can take up to *10^100 years.* As black holes evaporate, they will release their energy into the universe, but this energy will be so diffuse that it will have no meaningful impact. The universe will be a cold, dark, and lifeless expanse, populated only by the faint remnants of its former glory.

The Last Light: Echoes of the Big Bang

Even in this distant future, there will be one faint reminder of the universe's fiery birth: the cosmic microwave background (CMB). This afterglow of the Big Bang, now just 2.725 degrees above absolute zero, will continue to cool as the universe expands. Eventually, it will become so faint that it will be undetectable, leaving the universe in a state of perfect darkness.

In the final stages of the Big Freeze, the universe will be a vast, empty expanse, populated only by the remnants of its former self: black holes, neutron stars, and the faint glow of radiation left over from the Big Bang. Even these remnants will eventually decay, leaving behind a universe that is cold, dark, and lifeless. Time, as we understand it, will cease to have meaning. There will be no events, no change, and no processes to mark the passage of time. It will be a universe frozen in a state of eternal stillness.

The End of Time

The Big Freeze is not just the end of the universe as we know it—it is the end of time itself. In a cosmos that has reached maximum entropy, there will be no energy gradients, no processes to drive change, and no events to mark the passage of moments. Time, as we measure it today—through the ticking of clocks, the fusion of stars, or the beating of hearts—will dissolve into irrelevance. The universe will become a static tableau, frozen in a state of eternal stillness. It will be a cosmic graveyard, silent and unchanging, where nothing ever happens.

Time's Final Breath

Time, in physics, is inextricably linked to *change*. Without motion, decay, or transformation, the concept of time loses its meaning. Imagine a clock whose gears have rusted to a halt, its hands frozen in place. Now scale that image to the universe: galaxies no longer spin, particles no longer collide, and even the quantum fluctuations that once sparked virtual particles fade into silence. This is the fate of a universe in heat death, a state where entropy has triumphed, and all thermodynamic processes grind to a halt.

Theoretical models suggest this final state could arrive in 10^100 years or more—a timescale so vast it defies comprehension. By then, even the most resilient structures, like supermassive black holes, will have evaporated via Hawking radiation, leaving behind only a dilute sea of photons and subatomic particles, stretched to near-zero energy by the relentless expansion of space.

In this frozen epoch, there will be no witnesses to the universe's demise. No civilizations, no stars, not even atoms will remain. The very fabric of spacetime, once rippling with gravitational waves and quantum vibrations, will lie inert. The universe will forget its own history—no remnants of galaxies, no fossils of planets, no echoes of life. The cosmic story, which began in a blaze of light 13.8 billion years ago, will end in a whisper of radiation too feeble to detect.

Yet this bleak vision holds a paradoxical beauty. The Big Freeze is a testament to the universe's impermanence, a reminder that even the grandest structures—galaxy clusters, black holes, and the laws of physics themselves—are transient. It forces us to confront the fragility of existence and the inevitability of decay.

Philosophical Paradoxes

The end of time raises profound questions:

- *What is the purpose of a universe destined for oblivion?*

- *Does meaning derive from permanence, or from the fleeting moments of order we create?*
- *If time ceases to exist, does the concept of "end" even hold meaning?*

Philosophers and scientists have grappled with these questions for centuries. The ancient Stoics saw the universe as cyclical, perpetually reborn in fire (*ekpyrosis*). Modern cosmology, however, paints a linear timeline—one that begins with the Big Bang and culminates in the Big Freeze. Yet both visions share a haunting truth: *all things must pass.*

Though the universe's fate is sealed, humanity's quest to understand it is anything but futile. Our curiosity—the drive to map galaxies, decode dark energy, and ponder entropy—defines us. In a universe governed by entropy, life itself is a rebellion against disorder. For a brief flicker of cosmic time, we create art, build civilizations, and seek answers to questions older than stars.

The poet T.S. Eliot wrote, "This is the way the world ends / Not with a bang but a whimper." The Big Freeze embodies this whimper—a slow fade into darkness. Yet in that fade, there is a lesson: *the universe's impermanence makes every moment of its existence extraordinary.*

A Glimmer of Hope?

Some theories suggest ways to cheat entropy—advanced civilizations harvesting energy from black holes, or creating "pocket universes" to escape heat death. But these are speculative, bordering on science fiction. The reality is starker: entropy always wins.

But perhaps there is solace in this inevitability. Just as a sunset's fleeting beauty gains poignancy from its transience, the universe's story gains majesty from its impermanence. The stars that die, the galaxies that fade—they are not tragedies, but chapters in a grand, ephemeral narrative.

The Legacy of the Universe

As we contemplate the Big Freeze, we are reminded of the universe's legacy—an epic saga spanning 13.8 billion years, written in stardust and sculpted by forces beyond imagination. This legacy is not merely a chronicle of cosmic events, but a tapestry woven from creation and destruction, order and chaos, life and death. From the primordial fire of the Big Bang to the silent expanse of heat death, the universe has been a theater of transformation, where hydrogen atoms forged in stellar furnaces become the building blocks of planets, oceans, and minds capable of contemplating their own origins.

A Story Written in Starlight

The universe's legacy is etched into every atom in our bodies. The carbon in our cells, the iron in our blood, and the oxygen we breathe were all forged in the hearts of dying stars. Supernovae, the explosive deaths of massive stars, scattered these elements across the cosmos, seeding future generations of stars, planets, and life. Even the gold in a wedding ring or the uranium in a reactor core is a relic of neutron star collisions—cataclysmic events that lit up the darkness of space billions of years ago. We are, quite literally, made of stardust, bound to the universe's history in the most intimate way possible.

The Human Chapter

In this grand narrative, humanity occupies a fleeting but extraordinary chapter. For a mere blink in cosmic time—around 0.003% of the universe's existence—we have gazed upward, decoded the laws of physics, and sent probes like *Voyager 1* and *JWST* to the edges of the solar system and beyond. Our curiosity has mapped the cosmic microwave background, detected gravitational waves, and glimpsed black holes. These achievements are not just scientific milestones; they are acts of defiance against entropy, proof that life can briefly impose order on chaos and extract meaning from the void.

Entropy and the Art of Existence

The universe's inevitable march toward heat death underscores a profound irony: *entropy fuels creativity.* Stars shine because they fight against gravitational collapse. Life thrives by maintaining delicate energy gradients. Human culture—our art, science, and stories—is a rebellion against disorder. Consider the *Golden Record* aboard Voyager, carrying the sounds of Earth into interstellar space. It is a time capsule, a whisper of beauty hurled into the cosmic dark, destined to outlast our planet, our star, and perhaps even our galaxy. In a universe sliding toward equilibrium, such acts are luminous sparks of meaning.

The universe's legacy is also a symphony of unanswered questions. Dark matter's invisible grip, dark energy's mysterious push, the enigma of quantum gravity—these mysteries remind us that the story is far from over. Projects like the *Event Horizon Telescope,* which captured the first image of a black hole, and experiments probing the quantum foam of spacetime hint that we are on the cusp of revelations that could rewrite our understanding of reality. Each discovery peels back a layer of the cosmic onion, revealing deeper wonders and fiercer puzzles.

A Future Written in Shadows

Even as the universe expands toward icy stillness, its legacy is not one of despair, but of awe. The heat death is not an end, but a phase—one that future civilizations (or their AI descendants) might navigate in ways we can't yet fathom. Could they harvest energy from evaporating black holes? Create pocket universes? Or encode their consciousness into photons adrift in the void? These speculations, once the realm of science fiction, are now serious topics in cosmology and philosophy. They remind us that the universe's story is limited only by the laws of physics—and we've only just begun to read the fine print.

The universe's impermanence is its greatest gift. Like cherry blossoms that bloom brilliantly for a week or a sand mandala painstakingly crafted only to be swept away, the cosmos derives its beauty from its fragility. The Milky Way's spiral arms, Saturn's rings, and Earth's biosphere are masterpieces of ephemeral art. Their fleeting existence urges us to cherish the here and now—to marvel at a sunset, decode a supernova's light, or ponder the quantum vacuum—not in spite of their impermanence, but *because* of it.

Chapter 4: The Big Rip – The End of Everything

Butterfly emerges from stellar demise in planetary nebula NGC 6302 (Credits: NASA, ESA and the Hubble SM4 ERO Team)

Imagine a universe where galaxies are shredded, stars are torn apart, and even atoms are ripped to pieces—not by black holes or colliding galaxies, but by the fabric of spacetime itself unraveling. This is the *Big Rip*, a cataclysmic endgame where dark energy grows so powerful that it overcomes all forces binding the universe together. It's a scenario more violent than the Big Crunch and more sudden than the Big Freeze—a cosmic apocalypse where spacetime itself becomes the executioner.

The Big Rip hypothesis emerged in the early 2000s, born from a chilling question: *What if dark energy isn't just constant, but grows stronger over time?* If true, the universe wouldn't just expand—it would self-destruct in a crescendo of exponential acceleration.

Phantom Energy: The Universe's Silent Killer

The Big Rip hinges on a hypothetical form of dark energy called phantom energy, a force so bizarre it defies the very laws of physics as we know them. Unlike ordinary dark energy—modeled by Einstein's cosmological constant—phantom energy doesn't just push the universe apart; it grows stronger as the cosmos expands. Imagine a fire that burns hotter the more it consumes, or a car that accelerates faster the longer it drives. Phantom energy is the cosmic equivalent: a runaway force that could tear the universe apart from the inside out.

The Equation of State

Phantom energy's behavior is defined by its equation of state, a parameter denoted by 'w'. This parameter describes the relationship between the pressure and density of dark energy:

- **$w=-1$**: This is the value for the cosmological constant, the simplest form of dark energy. It remains constant over time, driving the universe's accelerated expansion at a steady rate.
- **$w<-1$**: This is the realm of phantom energy. Here, the pressure becomes so negative that the energy density *increases* as the universe expands. It's like stretching a rubber band and watching it grow thicker instead of

thinner—a violation of our everyday intuition.

Current observations, such as those from the *Dark Energy Survey* and the *Planck satellite*, suggest that 'w' is very close to -1, with a slight margin of error. This leaves the door open for phantom energy's terrifying possibility. If 'w' is even slightly less than -1, the universe could be on a collision course with the Big Rip!

A Violation of Physics?

Phantom energy doesn't just challenge our understanding of cosmology—it upends fundamental physics. One of the cornerstones of modern physics is the *weak energy condition*, which states that energy density cannot become negative. Phantom energy violates this condition, leading to bizarre consequences:

- *Negative Energy Density*: As phantom energy grows, it creates regions of negative energy, which could destabilize spacetime itself.
- *Causality Breakdown*: In extreme cases, phantom energy could create closed timelike curves, allowing time travel and breaking causality.
- *Quantum Instabilities*: Phantom energy could trigger quantum fluctuations on cosmic scales, leading to unpredictable and potentially catastrophic effects.

These violations make phantom energy a controversial idea, but they also make it fascinating. It's a reminder that the universe is stranger than we can imagine, and that our current laws of physics may only be a small piece of the cosmic puzzle.

So far, there's no direct evidence for phantom energy, but the possibility lingers in the margins of observational data. For example:

- *Supernova Surveys*: Observations of distant Type Ia supernovae show that the universe's expansion is accelerating, but they can't yet rule out $w<-1$.

- *Baryon Acoustic Oscillations (BAO)*: These "cosmic rulers," imprinted in the distribution of galaxies, provide another way to measure dark energy's properties. Current BAO data also leaves room for phantom energy.
- *Cosmic Microwave Background (CMB)*: The CMB's subtle temperature fluctuations hint at the universe's composition, but they don't yet pin down 'w' with absolute precision.

Upcoming missions like the Euclid Space Telescope and the Vera C. Rubin Observatory aim to measure 'w' with unprecedented accuracy. If they find $w<-1$, it could confirm the existence of phantom energy—and the inevitability of the Big Rip.

Phantom energy presents a paradox: it's both a destroyer and a creator. On one hand, it could rip the universe apart, ending all structure and life. On the other hand, its existence would reveal new physics, opening doors to discoveries we can't yet imagine. It's a reminder that the universe is full of surprises, and that even the most terrifying possibilities can lead to profound insights.

Timeline to Armageddon

If phantom energy exists, the countdown to the Big Rip might look like this:

1. *Today to 100 Million Years From Now:* Galaxies beyond our Local Group vanish over the cosmic horizon as expansion outpaces light. The Milky Way and Andromeda merge into a solitary "Milkomeda" galaxy.
2. *~60 Million Years Before the Rip:* Expansion overpowers gravity. The Milky Way is torn apart—stars drift into isolation.
3. *3 Months Before the Rip*: Solar systems disintegrate. Planets detach from their stars.
4. *30 Minutes Before the Rip:* Stars and planets explode as electromagnetic forces fail.

5. *Final Moments:* Atomic nuclei are shredded. Spacetime fractures. The universe becomes a chaotic foam of particles and radiation.

This timeline isn't just theoretical—it's mathematically grounded in phantom energy's equations.

The Physics of Destruction

The Big Rip isn't merely a tale of expansion; it's the story of spacetime unraveling with an insatiable hunger, stretching faster than light itself. Yes, you read that correctly—faster than light. While Einstein's theory of relativity famously imposes a cosmic speed limit on objects within spacetime, the fabric of spacetime itself is under no such constraint. This distinction allows for one of the most chilling scenarios in cosmology: a universe where nothing—not galaxies, not stars, not even atoms—can escape being torn apart.

At the heart of this apocalyptic vision lies the concept of the *Hubble Horizon* , an invisible boundary beyond which galaxies recede from us at velocities exceeding the speed of light. In our current epoch, this horizon sits far away, encompassing vast swaths of the observable universe. But in a Big Rip scenario, something sinister happens—the Hubble Horizon begins to shrink. It doesn't just contract like a balloon losing air; it devours everything in its path. Imagine it as a tightening noose, pulling inward inexorably, erasing the cosmos piece by piece. Galaxies once thought eternal are first to go, their distant lights extinguished as they cross the ever-narrowing horizon. Then come clusters, then individual stars—and eventually, when the horizon collapses down to subatomic scales, even the bonds holding protons and neutrons together will snap. The very essence of matter dissolves into the void, leaving behind… what? A featureless expanse of emptiness? Perhaps. Or perhaps something stranger still—a universe unmoored from all known laws of physics!

But before we reach that final cataclysmic moment, there's another horrifying phase: *gravitational decoupling* . Gravity, the gentle yet relentless force that has sculpted the cosmos since time immemorial, is rendered powerless against the accelerating tide of destruction. Picture gravity as the glue binding the universe together—the scaffolding upon which galaxies, solar systems, and planets are built. Now imagine that glue evaporating before your eyes. As dark energy ramps up its dominance, structures that have endured for billions of years begin to dissolve like sandcastles caught in a rising tsunami. Galaxy clusters, those majestic archipelagos of starlit islands, drift apart first, their mutual gravitational embrace too feeble to resist the pull of expansion. Next, individual galaxies lose cohesion, their spiral arms flung outward like ribbons caught in a hurricane. Closer to home, the delicate balance of forces keeping our solar system intact falters. Planets spiral out of orbit, moons careen wildly through space, and Earth itself may be stripped bare, its oceans and atmosphere ripped away like tissue paper in a gale.

And yet, the true terror of the Big Rip lies not in these grandiose acts of celestial demolition—but in the intimate, inevitable fate awaiting every atom in existence. When the expansion reaches its crescendo, the forces that hold atomic nuclei together—the strong nuclear force, nature's mightiest tether—will succumb to the overwhelming power of dark energy. Protons and neutrons will disintegrate, electrons will scatter, and the building blocks of reality will dissolve into a sea of fundamental particles adrift in infinite nothingness. At this point, the universe ceases to resemble anything we recognize. Space and time, those foundational pillars of existence, become meaningless concepts in a realm where causality itself has been shredded.

What makes the Big Rip so profoundly unsettling is how methodical it feels. Unlike other end-of-the-world scenarios—like a fiery supernova or a sudden collapse into a black hole—the Big Rip unfolds slowly, almost cruelly. It gives you time to watch, to comprehend, to despair. You see the writing on the wall long before the final curtain falls. Each stage of dissolution serves as a grim reminder of humanity's insignificance in the face of cosmic inevitability. And yet, paradoxically, it also underscores the beauty and resilience of the universe we inhabit today. For now, gravity holds firm. Stars shine. Life exists. We are here, fragile but defiant, pondering the ultimate fate of creation while standing atop a pale blue dot suspended in the vastness of spacetime.

Could We Survive?

The answer, as much as it pains me to admit, is a resounding *no* . Not even the most advanced technology or indestructible material could shield humanity—or any form of life—from the unrelenting fury of the Big Rip. Consider this: the very fabric of reality would be torn apart at its seams. Spacetime itself—the stage upon which all existence unfolds—would stretch and splinter until it unravels completely. No bunker, no spaceship, no Dyson sphere could protect us from an enemy that doesn't attack structures but dismantles the rules governing those structures in the first place.

Even black holes—the titans of the cosmos, regions so dense that not even light can escape their grasp—would meet a grim end. These celestial behemoths, often considered eternal sentinels of the universe, would find themselves powerless against the encroaching tide of destruction. As the Hubble Horizon contracts, the event horizons of black holes—the boundaries beyond which nothing returns—would disintegrate. Without these defining edges, black holes would lose their identity, their mass spilling out into the void like sand leaking through cracked fingers. They would evaporate not due to Hawking radiation, as some eventually might over trillions of years, but because the very concept of "space" within which they exist would cease to hold meaning. In the face of such cosmic annihilation, survival becomes not just improbable—it becomes inconceivable.

Yet, the implications of the Big Rip extend far beyond mere physical survival. This scenario forces us to confront one of the most haunting philosophical questions ever posed: *If the universe is doomed to vanish without a trace, does anything we do matter?* At first glance, the question seems almost paralyzing. Why strive for progress, create art, build civilizations, or seek knowledge if everything will ultimately be erased? If the stars themselves are destined to fade into oblivion, what purpose do our fleeting lives serve?

The Limits of Technology

Human ingenuity has taken us to the Moon, probed the depths of the ocean, and unlocked the secrets of the atom. But the Big Rip is a force beyond our control. Consider the following:

- *Energy Requirements:* To counteract the Rip, we'd need energy on a scale that dwarfs the output of entire galaxies. Even harnessing the power of a Type III civilization (one capable of controlling the energy of an entire galaxy) would be futile against phantom energy's exponential growth.
- *Structural Integrity:* No material, not even graphene or neutron star crust, could withstand the stretching of spacetime itself. The Rip doesn't just destroy objects—it destroys the very framework in which they exist.
- *Time Constraints:* If the Rip occurs, the final moments would be measured in minutes or seconds. There would be no time to escape, no time to adapt, and no time to say goodbye.

A Haunting Philosophical Question

The inevitability of the Big Rip raises a profound and unsettling question: *If the universe is doomed to vanish without a trace, does anything we do matter?*

- *The Absurdist Perspective*: Philosophers like Albert Camus might argue that the universe's meaninglessness is precisely what makes life meaningful. In a cosmos destined for oblivion, every act of creation, every moment of joy, and every act of kindness becomes a rebellion against the void.
- *The Legacy of Knowledge*: Even if the universe is fated to end, the pursuit of knowledge leaves a legacy. The equations of Einstein, the art of Van Gogh, and the music of Beethoven are not just products of their time—they are timeless expressions of human creativity and curiosity.
- *The Cosmic Perspective*: The Big Rip reminds us that we are part of

something vast and mysterious. Our existence, fleeting as it may be, is a testament to the universe's ability to create complexity and consciousness. In that sense, we are not just observers of the cosmos—we are its storytellers.

While survival is impossible, the Big Rip doesn't render life meaningless. On the contrary, it underscores the preciousness of the present. Every sunrise, every act of love, and every scientific discovery becomes a fleeting spark of beauty in an otherwise indifferent universe.

Moreover, the Big Rip is not a certainty. It's one of many possible futures, each dependent on the nature of dark energy. If ww (the equation of state parameter) is exactly -1, the universe will expand forever without tearing itself apart. If ww is greater than -1, the expansion will slow, potentially leading to a Big Crunch or a cyclic universe. The uncertainty itself is a call to action—a reminder that the story of the universe is still being written, and we have a role to play in shaping its narrative.

As the poet Mary Oliver once asked, *"Tell me, what is it you plan to do with your one wild and precious life?"* In a universe that may one day vanish without a trace, the answer lies not in survival, but in the beauty we create, the connections we forge, and the questions we dare to ask.

A Glimmer of Doubt

While the Big Rip is a compelling and dramatic scenario, not all cosmologists are convinced it's our universe's fate. Critics of the hypothesis point to gaps in our understanding and the lack of definitive evidence for phantom energy. The Big Rip remains a tantalizing possibility, but it's far from a foregone conclusion. Let's explore the key arguments against it—and why the mystery of dark energy continues to keep scientists on the edge of their seats.

Observational Limits: The Data Isn't In Yet

Current observations of the universe's expansion, while precise, still leave room for interpretation. The equation of state parameter 'w', which determines the behavior of dark energy, is currently measured to be very close to -1, but with a margin of error that doesn't rule out $w<-1$.

- *Supernova Surveys:* Observations of Type Ia supernovae, which serve as cosmic mile markers, have been instrumental in measuring the universe's expansion. However, they haven't yet provided conclusive evidence for phantom energy. The *Dark Energy Survey* and the upcoming *Vera C. Rubin Observatory* aim to refine these measurements, but for now, the data remains ambiguous.
- *Baryon Acoustic Oscillations (BAO)*: These "cosmic rulers," imprinted in the distribution of galaxies, provide another way to measure dark energy's properties. Current BAO data also leaves room for $w<-1w<-1$, but it doesn't confirm it.
- *Cosmic Microwave Background (CMB)*: The CMB's subtle temperature fluctuations hint at the universe's composition, but they don't yet pin down ww with absolute precision.

The *Euclid Space Telescope*, launched in 2024, could change everything. Designed to map the distribution of galaxies and measure the universe's expansion with unprecedented accuracy, Euclid may finally answer the question: *Is phantom energy real?*

Quantum Gravity Unknowns: The Planck Scale Wildcard

Even if phantom energy exists, there's another layer of uncertainty: quantum gravity. At the Planck scale—the smallest possible scale in physics—our current understanding of spacetime breaks down. General relativity, which describes the universe on large scales, and quantum mechanics, which governs the very small, are fundamentally incompatible. This means that as the Big Rip approaches, quantum effects could come into play, potentially halting the Rip or altering its outcome.

- *Loop Quantum Gravity*: This theory suggests that spacetime is not continuous but made up of discrete "loops." At extremely small scales, these loops could resist infinite stretching, preventing the Big Rip.
- *String Theory*: In this framework, spacetime is a higher-dimensional construct, and the Big Rip might be averted by interactions in these extra dimensions.
- *Quantum Fluctuations*: At the Planck scale, spacetime itself might become a seething foam of quantum fluctuations, potentially stabilizing the universe against phantom energy's destructive force.

These theories are speculative, but they highlight the limits of our current knowledge. The Big Rip assumes that dark energy behaves the same way at all scales—but quantum gravity could rewrite the rules entirely.

The Sword of Damocles

Despite these doubts, the possibility of the Big Rip remains—a cosmic sword of Damocles hanging over the universe. It's a reminder that our understanding of dark energy is still in its infancy, and that the universe is full of surprises.

- *The Multiverse Angle*: Some theories suggest that phantom energy could trigger the creation of new universes, turning the Big Rip into a cosmic reset button rather than an end.
- *The Role of Dark Matter*: If dark energy interacts with dark matter in unexpected ways, it could alter the universe's fate. For example, a decaying dark energy field might transition into a form that doesn't lead to a Rip.

- *The Anthropic Principle*: Some physicists argue that if the Big Rip were inevitable, we wouldn't be here to observe it. The fact that we exist might imply that ww is not less than -1—or that our universe is part of a multiverse where only "Rip-free" universes can support life.

The debate over the Big Rip is a microcosm of cosmology itself—a field defined by uncertainty, curiosity, and the thrill of discovery. Whether or not the Rip is real, it challenges us to push the boundaries of our knowledge, to question our assumptions, and to explore the deepest mysteries of the cosmos.

The Big Rip and Human Legacy

If the Big Rip is real, it doesn't just spell the end of the universe—it forces us to confront the impermanence of everything we hold dear. Galaxies, stars, planets, and even the fabric of spacetime itself would be torn apart, leaving no trace of the cosmos as we know it. In the face of such oblivion, what becomes of human legacy? What does it mean to create, to explore, or to love in a universe destined to vanish without a trace?

The Urgency to Explore and Document

The possibility of the Big Rip amplifies the urgency to explore and document the universe. If the cosmos is fated to end in a violent rupture, then every moment of discovery becomes a race against time. Projects like the Lunar Library and Voyager's Golden Record take on new poignancy—they are not just messages to future civilizations, but time capsules meant to outlast humanity itself.

- *The Lunar Library*: A nickel-etched archive buried on the Moon, the Lunar Library contains millions of pages of human knowledge, from the

works of Shakespeare to the schematics of the Apollo missions. It's a backup of human civilization, designed to endure for billions of years.
- *Voyager's Golden Record*: Launched in 1977, the Voyager spacecraft carry golden records containing sounds and images of Earth—a greeting to any extraterrestrial beings who might find them. Even if the Big Rip occurs, these records could drift through the void for eons, a faint whisper of humanity's existence.
- *The Arch Mission Foundation*: This nonprofit organization aims to create a "backup of planet Earth," storing human knowledge in durable formats like synthetic DNA and nanoscale etchings. These archives are designed to survive cosmic disasters, including the Big Rip.

These projects are more than just scientific endeavors—they are acts of defiance against entropy, proof that even in the face of annihilation, humanity strives to leave a mark.

Redefining Meaning in a Fleeting Universe

The Big Rip challenges us to redefine what it means to live a meaningful life. In a universe where nothing lasts, the act of creation becomes its own purpose. A poem, a symphony, or a child's laughter gains weight not from permanence, but from its fleeting brilliance.

- *The Beauty of Transience*: The Japanese concept of *mono no aware*—the awareness of impermanence—captures this idea perfectly. It's the bittersweet recognition that all things must pass, and that their fleeting nature is what makes them beautiful. A cherry blossom is precious because it blooms for only a week; a sunset is breathtaking because it lasts only minutes. In a universe hurtling toward the Big Rip, every moment becomes a cherry blossom, every act of creation a sunset.
- *Art as Rebellion*: In the face of cosmic oblivion, art becomes a form of

rebellion. Whether it's a painting, a song, or a scientific theory, every act of creation is a defiance of entropy, a spark of order in a universe sliding toward chaos.

- *The Legacy of Curiosity*: The Big Rip reminds us that the pursuit of knowledge is its own reward. Even if the universe is fated to end, the questions we ask and the discoveries we make are a testament to humanity's boundless curiosity.

The Last Question

The Big Rip forces us to confront the ultimate mystery: *Why is there something rather than nothing?* If the universe can end so violently, why did it begin at all? These questions have no answers—yet. But in asking them, we join a lineage of thinkers stretching back to the first humans who gazed at the stars and wondered. The Big Rip may be the universe's final act, but the story of curiosity—of life daring to understand its place in the cosmos—is one that can never be torn apart.

The question of why the universe exists at all is one of the oldest and most profound in human history. Philosophers, theologians, and scientists have grappled with it for millennia, yet it remains unanswered. The Big Rip adds a new layer to this mystery: if the universe is destined to end in such a violent and final way, what was the point of its existence?

Some scientists argue that the universe exists because we are here to observe it. In other words, the conditions necessary for life—the precise values of physical constants, the existence of dark energy, and the balance of forces—are what make the universe possible. But this raises another question: why do these conditions exist in the first place?

One possible answer is that our universe is just one of many in a vast multiverse. In this view, the Big Rip is just one of countless possible fates, each playing out in a different universe. But this only pushes the question back a step: why does the multiverse exist?

While the Big Rip is a terrifying possibility, it's also a call to action. It reminds us that the universe is a place of infinite mystery, and that the journey to understand it is far from over. Whether or not the Rip occurs, the act of exploration—of pushing the boundaries of knowledge and creativity—is what makes life worth living.

In the end, the Big Rip is not just a scientific hypothesis—it's a mirror reflecting our deepest fears and aspirations. It challenges us to confront the fragility of existence, to find meaning in the face of impermanence, and to cherish the brief, brilliant moment we have in the cosmic spotlight.

As the poet Rainer Maria Rilke once wrote, *"Be patient toward all that is unsolved in your heart and try to love the questions themselves."* In a universe that may one day vanish without a trace, the questions we ask—and the beauty we create—are what truly matter.

Chapter 5: The Multiverse – Escaping the End?

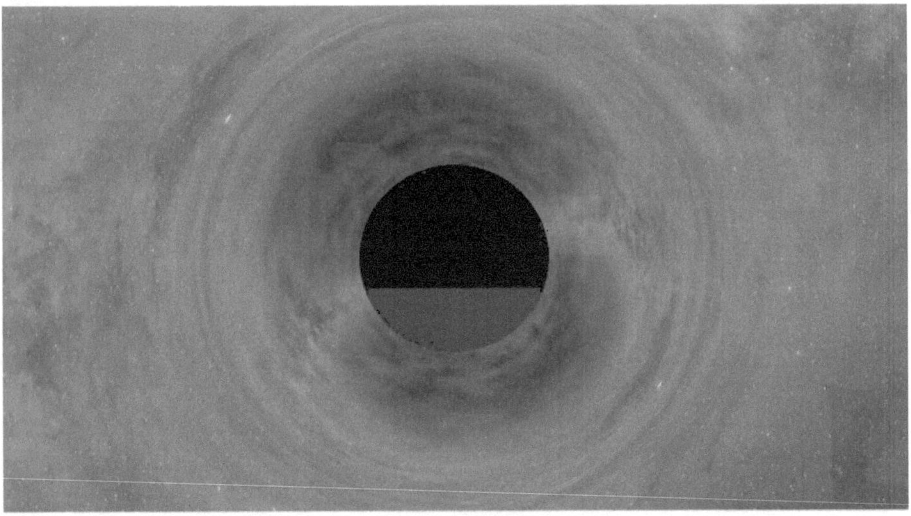

Isolated Black Hole Visualization (Credit: NASA)

What if our universe isn't the only one? What if, beyond the reach of our telescopes and particle colliders, there exist countless other universes—each with its own laws of physics, its own stars and galaxies, and perhaps its own civilizations grappling with their own cosmic fates? This is the multiverse, a speculative but tantalizing idea that offers a radical escape from the universe's doom. If the Big Rip, Big Crunch, or Big Freeze is inevitable, could life, information, or even consciousness survive by fleeing to another universe?

The multiverse hypothesis isn't science fiction—it's a serious scientific proposal rooted in quantum mechanics, string theory, and cosmic inflation. While unproven, it challenges us to rethink the very nature of reality and our place within it.

A Universe of Universes

The idea of a multiverse isn't new. Ancient philosophers like Democritus speculated about infinite worlds, and modern science has given the concept a rigorous framework. The multiverse emerges naturally from several well-established theories:

- *Quantum Mechanics:* The "many-worlds interpretation" suggests that every quantum event spawns a new universe. For example, when you flip a coin, one universe sees it land heads, and another sees tails. Extend this to cosmic scales, and there could be universes where dark energy behaves differently, averting the Big Rip or triggering a Big Crunch.
- *Cosmic Inflation:* The rapid expansion of the early universe may still be happening in distant regions, creating an endless froth of "bubble universes." Each bubble has its own physical laws, born from random fluctuations in the inflaton field.
- *String Theory:* This framework proposes that our universe is a 3D "brane" floating in a higher-dimensional "bulk." Other branes—parallel universes—could exist mere millimeters away, invisible to us.

The multiverse has captured the imagination of storytellers, from comic books to blockbuster movies. Marvel's *Spider-Man: Into the Spider-Verse* and *Doctor Strange in the Multiverse of Madness* explore the idea of infinite realities, each with its own version of heroes and villains. The TV show *Rick and Morty* takes a darker, more absurdist approach, with its titular characters hopping between universes to escape consequences or find better versions of their lives.

But how much of this is grounded in science? While these stories are entertaining, they often take creative liberties. For example, the idea of "jumping" between universes with a portal gun or a spell is pure fiction—at least for now. The real multiverse, if it exists, would likely be far more alien and inaccessible.

The Philosophical Implications

The multiverse hypothesis raises profound philosophical questions:

- *Are We Alone?*: If there are infinite universes, does that mean there are infinite versions of us? And if so, what does that say about free will and individuality?
- *The Meaning of Life*: In a multiverse, does life have a purpose, or is it just a statistical inevitability?
- *The Nature of Reality*: If our universe is just one of many, does that make it less "real"? Or does it make the multiverse itself the ultimate reality?

Not all scientists are convinced. Critics argue that the multiverse is untestable and therefore unscientific. After all, if we can't observe or interact with other universes, how can we prove they exist?

The Quantum Multiverse: Infinite Possibilities

Tidal Disruption Events (Credit: NASA)

In the quantum multiverse, every decision, every random quantum event, spawns a new universe. This idea, popularized by the "many-worlds interpretation" of quantum mechanics, suggests that every possible outcome of every event exists in a branching network of parallel realities. It's a dizzying concept: every choice you've ever made—or didn't make—has played out in some other universe.

Schrödinger's Cat Revisited

The classic thought experiment of Schrödinger's cat illustrates the strange nature of quantum mechanics. In one universe, the cat lives; in another, it dies. Extend this to cosmic scales, and the implications are staggering. There could be universes where:

- *Dark Energy Behaves Differently*: In some universes, dark energy might have a value of $w > -1$, preventing the Big Rip and allowing the universe to expand forever. In others, $w < -1$ might lead to a Big Rip, tearing spacetime apart.

- *Gravity Wins*: In some universes, gravity might dominate; pulling everything back together in a fiery Big Crunch.
- *Life Evolves Differently*: In some universes, the laws of physics might allow for entirely new forms of life, or even prevent life from arising at all.

Escaping the End

If advanced civilizations could harness quantum tunneling or traverse wormholes, they might leap into a neighboring universe before their own meets its demise. But this remains speculative, requiring physics far beyond our current understanding.

- *Quantum Tunneling*: This phenomenon allows particles to pass through barriers that would be insurmountable under classical physics. In theory, a civilization could use quantum tunneling to "jump" into a parallel universe. However, scaling this up to macroscopic objects—let alone entire civilizations—is a monumental challenge.
- *Wormholes*: These hypothetical tunnels through spacetime could connect distant regions of the universe—or even different universes. But wormholes are highly unstable and would require exotic matter with negative energy to keep them open.

The Inflationary Multiverse: Bubble Universes

Cosmic inflation—the rapid expansion of the early universe—may still be happening in distant regions of space, creating an endless froth of "bubble universes." Each bubble has its own physical laws, born from random fluctuations in the inflaton field. This idea, known as the inflationary multiverse, suggests that our universe is just one of countless others, each with its own unique properties and fate.

According to theorists like *Alan Guth* and *Andrei Linde*, inflation never truly stops. Instead, it continues eternally in some regions, spawning new universes like bubbles in a cosmic champagne glass.

- *The Inflaton Field*: This hypothetical field drove the rapid expansion of the early universe. In some regions, the field's energy drops, causing inflation to end and a new universe to form. In others, inflation continues, creating an endless cascade of bubble universes.
- *Diverse Laws of Physics*: Each bubble universe could have different physical constants, dimensions, and even fundamental forces. In some, gravity might be stronger; in others, electromagnetism might dominate. This diversity could explain why our universe seems so finely tuned for life—we just happen to be in a bubble where the conditions are right.

A Lifeline for Life?

If our universe is doomed, could life migrate to a younger, more hospitable bubble? The idea is tantalizing but fraught with challenges:

- *Bubble Walls*: The boundaries between bubble universes might be lethal membranes of high-energy physics. Crossing them could be like trying to swim through a wall of plasma.
- *Cosmic Distances*: The distances between bubble universes could be unimaginably vast. Even if we could survive the journey, the energy required to traverse such distances would dwarf anything achievable with current technology.
- *Unknown Conditions*: We have no idea what conditions exist in other bubble universes. They might be inhospitable to life, or they might have already met their own cosmic fates.

The Brane Multiverse: Universes Next Door

String theory, one of the most ambitious frameworks in modern physics, proposes a radical idea: our universe is a 3D "brane" (short for membrane) floating in a higher-dimensional "bulk." This bulk could be home to countless other branes—parallel universes—each with its own physical laws, stars, and perhaps even life. These universes might be mere millimeters away from us in higher dimensions, yet completely invisible and inaccessible due to the limitations of our 3D perception.

One of the most dramatic implications of the brane multiverse is the possibility of *brane collisions*. Some models, such as the *Ekpyrotic Universe* theory, suggest that the Big Bang itself was caused by a collision between two branes. Imagine two sheets of paper floating in a higher-dimensional space, drifting closer and closer until they slam into each other. The energy released in such a collision could ignite a new universe, complete with its own laws of physics and initial conditions.

- *A Cyclic Universe*: If brane collisions are periodic, the universe could undergo endless cycles of creation and destruction. Each collision would reset the cosmic clock, offering a fresh start. This idea echoes ancient philosophical concepts like the Hindu notion of *pralaya* (cosmic dissolution) and the Stoic idea of *ekpyrosis* (universal conflagration).
- *Evidence for Collisions*: While direct evidence for brane collisions is lacking, some theorists have proposed that anomalies in the cosmic microwave background (CMB) could be "bruises" left by such events. These would appear as subtle temperature variations or patterns in the CMB, though none have been conclusively identified yet.

Interdimensional Travel: Escaping the Doomed Brane

If our brane is destined for a Big Rip, Big Crunch, or heat death, could advanced civilizations escape by moving into the bulk? The idea of interdimensional travel is a staple of science fiction—think *Stranger Things*' Upside Down or *Doctor Who*'s TARDIS—but could it ever be scientifically plausible?

- *The Energy Barrier*: Moving between branes would require overcoming immense energy barriers. The forces binding our brane to the bulk are unimaginably strong, and the energy required to break free would dwarf anything achievable with current technology. However, a civilization that survives for billions of years might develop the means to harness such energy.
- *The Bulk as a Lifeline*: If life could encode itself into higher-dimensional forms—perhaps as patterns of energy or information—it might traverse the bulk and colonize a neighboring brane. This would be the ultimate act of cosmic survival, a Hail Mary pass to escape the end of our universe.

Pop Culture Parallels

The brane multiverse has captured the imagination of storytellers and filmmakers, though often with a generous dose of artistic license.

- *Marvel's Multiverse*: In *Spider-Man: No Way Home* and *Doctor Strange in the Multiverse of Madness*, the multiverse is a playground of infinite possibilities, where every decision spawns a new reality. While entertaining, these depictions gloss over the scientific challenges of interdimensional travel.
- *Interstellar*: Christopher Nolan's film touches on higher dimensions through the tesseract, a 5D construct that allows Cooper to communicate across time and space. While speculative, it hints at the mind-bending

possibilities of higher-dimensional physics.

- *The Matrix*: While not explicitly about branes, the idea of a simulated reality raises similar questions about the nature of existence and the possibility of escaping to a "higher" reality.

The Scientific Credibility

While the brane multiverse is a compelling idea, it remains speculative. String theory, which underpins the concept, has yet to produce testable predictions. Critics argue that without empirical evidence, the brane multiverse is more philosophy than science.

- *The Challenge of Testing*: Because higher dimensions are, by definition, inaccessible to us, proving the existence of branes is incredibly difficult. Some physicists hope that experiments at the *Large Hadron Collider (LHC)* might detect evidence of extra dimensions, such as microscopic black holes or gravitons leaking into the bulk, but so far, no such evidence has been found.
- *The Role of Mathematics*: String theory relies heavily on advanced mathematics, leading some to question whether it describes reality or is simply an elegant abstraction. As physicist Richard Feynman once quipped, "If it disagrees with experiment, it's wrong."

The brane multiverse offers a tantalizing possibility: that the end of our universe might not be the end of everything. If advanced civilizations can master higher-dimensional physics, they might find a way to escape their dying brane and start anew in a neighboring universe.

The Holographic Multiverse: Reality as an Illusion

The holographic principle is one of the most mind-bending ideas in modern physics. It suggests that our 3D universe—with its galaxies, stars, and planets—is actually a projection of information encoded on a 2D surface, like a cosmic DVD. This isn't just a metaphor; it's a mathematical framework that emerges from the study of black holes and quantum gravity. Extend this idea, and multiple universes could exist as overlapping holograms on a shared boundary, each one a unique projection of information.

Black Hole Universes: Portals to New Realities

One of the most intriguing implications of the holographic principle is the idea that black holes might be gateways to other universes. According to some theories, the information that falls into a black hole isn't destroyed—it's encoded on the black hole's event horizon, the 2D boundary that marks the point of no return.

- *Holographic Birth*: Some physicists, like Leonard Susskind and Juan Maldacena, propose that the interior of a black hole could be a new universe, with its own spacetime and physical laws. In this scenario, our universe might itself be the interior of a black hole in a higher-dimensional parent universe.
- *Escaping the End*: If life could encode itself into information—perhaps as a form of digital consciousness—it might cross the event horizon and enter a new holographic universe. This would be the ultimate survival strategy: not fleeing to another planet or galaxy, but transcending the very fabric of spacetime.

Data Survival: The Cosmic Backup

Even if our universe meets its end in a Big Rip, Big Crunch, or heat death, the "source code" of reality might persist on the holographic boundary. This boundary, often compared to a cosmic hard drive, could store all the information needed to reconstruct the universe—or even create a new one.

- *Rebooting Reality*: In theory, a sufficiently advanced civilization could access this information and use it to reboot the universe. This wouldn't be a resurrection of the old cosmos, but the creation of a new one, perhaps with different physical constants or laws.
- *The Library of Babel*: Imagine a cosmic archive where every possible universe is encoded as a unique set of data. Our universe might be just one entry in this infinite library, waiting to be accessed or rewritten.

Philosophical Implications: Is Reality Real?

The holographic principle challenges our most basic assumptions about reality. If the universe is a projection, then what we perceive as solid, tangible objects might be nothing more than patterns of information.

- *The Matrix Revisited*: The idea of a simulated or projected reality isn't new—it's been explored in films like *The Matrix* and *Inception*. But the holographic principle takes this a step further, suggesting that even the laws of physics might be emergent properties of information.
- *The Nature of Consciousness*: If reality is a hologram, then consciousness itself might be a form of information processing. This raises profound questions about the nature of self, free will, and the meaning of existence.

Scientific Challenges: Can We Test the Holographic Principle?

While the holographic principle is mathematically elegant, it's also notoriously difficult to test. The energy scales involved are far beyond anything we can probe with current technology.

- *Quantum Gravity Experiments*: Some physicists hope to find evidence of the holographic principle in experiments that probe the quantum nature of spacetime, such as those involving gravitational waves or high-energy particle collisions.
- *Cosmic Microwave Background*: The CMB, the afterglow of the Big Bang, might contain subtle imprints of the holographic boundary. Analyzing these imprints could provide clues about the true nature of reality.

The holographic principle is a reminder that the universe is stranger than we can imagine. It suggests that reality is not what it seems—that the solid ground beneath our feet and the stars above our heads might be nothing more than a grand illusion.

The Fermi Paradox Twist

The multiverse adds a new layer to the Fermi Paradox: If countless universes exist, why haven't we detected any signs of cross-universe civilizations? The Fermi Paradox, which asks why we haven't found evidence of extraterrestrial life despite the vastness of our own universe, becomes even more perplexing when you consider the possibility of an infinite multiverse. If there are infinite universes with infinite possibilities, where is everyone?

Dimensional Barriers: The Walls Between Worlds

One possibility is that dimensional barriers make travel between universes physically impossible. Even if other universes exist, they might be separated by insurmountable distances or higher-dimensional boundaries that no technology, no matter how advanced, can cross.

- *The Bulk Problem*: In string theory, our universe is a 3D brane floating in a higher-dimensional "bulk." If other branes exist, they might be millimeters away in these extra dimensions—but those millimeters could be as impassable as a brick wall.
- *Energy Requirements*: Crossing between universes might require energy levels far beyond anything achievable, even for a Type III civilization (one capable of harnessing the energy of an entire galaxy).

Silent Neighbors: The Invisible Multiverse

Another possibility is that civilizations in other universes are silent neighbors—either by choice or necessity.

- *Undetectable Communications*: If other universes operate under different physical laws, their methods of communication might be completely alien to us. A signal sent via gravity waves in one universe might be invisible in another.
- *The Dark Forest Hypothesis*: Borrowing from Liu Cixin's *The Three-Body Problem*, civilizations might choose to remain silent to avoid detection. In a multiverse teeming with life, staying hidden could be the ultimate survival strategy.
- *The Great Silence*: It's also possible that civilizations capable of escaping their dying universes are so advanced that they've transcended physical forms, existing as pure information or energy. To us, they might be

invisible.

The Great Filter: A Cosmic Bottleneck

The Great Filter theory suggests that there's a nearly insurmountable barrier preventing civilizations from reaching a stage where they can escape their universe. This filter could occur at any point in a civilization's development:

- *Early Filters:* Civilizations might self-destruct through war, environmental collapse, or misuse of technology before they can develop the means to escape.
- *Late Filters*: Even advanced civilizations might find the physics of interdimensional travel too complex or the energy requirements too prohibitive.
- *The Ultimate Filter*: The ability to escape a dying universe might be so rare that no civilization has ever achieved it. If true, this would make our universe's fate all the more poignant.

Some theorists suggest that we haven't detected cross-universe civilizations because we're being deliberately isolated—like animals in a cosmic zoo. Advanced civilizations might be observing us from afar, choosing not to interfere until we reach a certain level of technological or ethical maturity.

- *The Prime Directive*: Borrowing from *Star Trek*, this idea posits that advanced civilizations follow a code of non-interference, allowing younger species to develop on their own.
- *The Simulation Argument*: If our universe is a simulation, the "operators" might be limiting our access to other universes to maintain the integrity of the experiment.

The Fermi Paradox and the Multiverse: A New Perspective

The multiverse doesn't just complicate the Fermi Paradox—it transforms it. If there are infinite universes, the absence of detectable life becomes even more puzzling. But it also offers new ways to think about our place in the cosmos:

- *Are We Alone in the Multiverse?*: If no civilizations have escaped their universes, it might suggest that the Great Filter is universal—a barrier that no species can overcome.
- *A Call to Action*: The Fermi Paradox, amplified by the multiverse, underscores the urgency of space exploration and technological advancement. If we want to survive the end of our universe, we need to start thinking beyond its boundaries.

The Fermi Paradox, in the context of the multiverse, is a reminder of how much we still have to learn. It challenges us to push the boundaries of science, to explore the unknown, and to ask questions that might never have answers.

The Simulation Hypothesis: A Digital Multiverse

What if our universe is a simulation running on a vast computer? This idea, known as the simulation hypothesis, suggests that everything we perceive as reality—stars, planets, consciousness, and even the laws of physics—is nothing more than lines of code in an unimaginably advanced program. In this scenario, the "multiverse" could simply be other simulations operating in the same substrate, each running its own version of reality.

Reboot Options: Escaping the End

If our universe is a simulation, its end—whether through the Big Rip, Big Crunch, or Big Freeze—might not be the final word. The operators of the simulation (whoever or whatever they are) could have contingency plans:

- *Rebooting the Simulation*: Just as a computer can restart after a crash, the simulation could be rebooted, resetting the universe to its initial conditions. This would erase all memory of the previous run, but it would also offer a fresh start.
- *Transferring Consciousness*: If the simulation's inhabitants have achieved a level of technological sophistication, they might encode their consciousness into data and transfer it to another simulation. This would be the ultimate form of survival, allowing life to persist even as its original universe ends.
- *Backup Universes*: The operators might create backup simulations, ensuring that if one universe fails, others can continue. This would make the multiverse a kind of cosmic redundancy system, where the end of one universe doesn't mean the end of all existence.

Ethical Quandaries: Rights in a Simulated Reality

The simulation hypothesis raises profound ethical questions. If we are living in a simulation, do we have rights? Could we plead with our creators to avoid deletion?

- *The Rights of Simulated Beings*: If a civilization creates a simulation populated with sentient beings, does it have a moral obligation to ensure their well-being? Philosophers like Nick Bostrom, who popularized the simulation hypothesis, argue that the ethical implications are immense. Deleting a simulation could be tantamount to genocide.

- *The Creator's Dilemma*: If the operators of the simulation are themselves simulated beings, they might be bound by their own ethical constraints. This creates an infinite regress of responsibility, where each layer of reality must grapple with the morality of creating and destroying universes.
- *The Illusion of Free Will*: If our actions are determined by the code of the simulation, do we truly have free will? And if not, can we be held morally responsible for our actions? These questions challenge our understanding of ethics, identity, and the nature of existence itself.

Pop Culture Parallels

The simulation hypothesis has captured the imagination of storytellers and filmmakers, becoming a staple of science fiction.

- *The Matrix*: Perhaps the most famous exploration of the simulation hypothesis, *The Matrix* depicts a world where humans are unknowingly trapped in a simulated reality. The film raises questions about the nature of reality, the ethics of simulation, and the possibility of escape.
- *Rick and Morty*: This animated series takes the multiverse to absurd extremes, with infinite parallel universes and countless versions of the same characters. While not strictly about simulation, it plays with the idea of alternate realities and the ethical dilemmas they create.
- *Westworld*: This series explores the ethics of creating sentient beings in a simulated environment, raising questions about the rights of artificial life and the responsibilities of creators.

Scientific Credibility: Is the Simulation Hypothesis Plausible?

While the simulation hypothesis is a fascinating thought experiment, its scientific credibility is hotly debated.

- *The Bostrom Trilemma*: Philosopher Nick Bostrom argues that at least one of the following must be true: (1) almost all civilizations go extinct before becoming technologically mature; (2) almost all technologically mature civilizations lose interest in creating simulations; or (3) we are almost certainly living in a simulation. While this argument is logically sound, it relies on assumptions that are impossible to test.
- *The Fermi Paradox Connection*: If advanced civilizations routinely create simulations, why haven't we detected any evidence of them? This adds another layer to the Fermi Paradox, suggesting that either simulations are undetectable or civilizations avoid creating them for ethical reasons.
- *The Limits of Computation*: Even if our universe is a simulation, the computational power required to simulate every particle and interaction would be beyond anything we can imagine. Some physicists argue that the universe's complexity makes the simulation hypothesis unlikely.

While the simulation hypothesis is speculative, it offers a glimmer of hope in the face of cosmic doom. If our universe is a simulation, its end might not be the end of everything. Consciousness, information, and even life itself might persist in other simulations or in the substrate of the original reality.

The Anthropic Shadow: Why We Might Never Know

If most universes in the multiverse are inhospitable to life, our existence might depend on being in a rare "Goldilocks" universe—one that's just right for the emergence of stars, planets, and conscious beings. This idea, known as the anthropic principle, offers a compelling explanation for why we find ourselves in a universe with stable atoms, dark energy, and other life-friendly conditions. But it also comes with a sobering implication: we're statistically unlikely to ever contact other universes, or even know for sure if they exist.

The anthropic principle suggests that the conditions of our universe aren't random—they're finely tuned to allow life to exist. For example:

- *The Cosmological Constant*: If dark energy were slightly stronger, galaxies would never have formed. If it were weaker, the universe might have collapsed in a Big Crunch long before life could emerge.
- *The Fine-Structure Constant*: This dimensionless number governs the strength of electromagnetic interactions. If it were even slightly different, atoms wouldn't form, and chemistry as we know it would be impossible.
- *The Mass of the Proton*: If protons were just 0.2% heavier, they would decay into neutrons, making stable matter impossible.

These "cosmic coincidences" suggest that our universe is one of the rare ones capable of supporting life. But if the multiverse is real, it's likely filled with countless universes where the constants are wrong, the laws of physics are hostile, and life is impossible.

The Anthropic Shadow

The anthropic principle casts a shadow over our ability to understand the multiverse. If most universes are lifeless, then the ones we can observe—ours included—are the exceptions, not the rule. This means that our observations are biased: we can only exist in a universe that allows us to exist.

- *The Fermi Paradox Revisited*: If the multiverse is teeming with universes, why haven't we detected any signs of life from other universes? The anthropic shadow suggests that most universes are barren, and the few that aren't are too far apart to communicate.
- *The Limits of Science*: If the multiverse is real, it might be fundamentally

unknowable. Universes with different physical laws might be inaccessible to us, and even if they're not, the sheer scale of the multiverse could make contact impossible.

The anthropic shadow reminds us that the universe (or multiverse) isn't obligated to make sense to us. It's a place of infinite mystery, where the rules of physics can vary wildly from one region to another. But it's also a place of infinite possibility, where life and consciousness can emerge against all odds.

The Legacy of Curiosity

The multiverse hypothesis, while unproven, is a testament to humanity's relentless creativity. It transforms the universe's end from a terminal prognosis into a cosmic challenge—a puzzle to be solved, a frontier to be crossed. The idea that there might be other universes, each with its own laws of physics and its own stories, is both humbling and exhilarating. It reminds us that the universe (or multiverse) is far stranger and more wondrous than we can imagine, and that the journey to understand it is far from over.

Every generation has its visionaries—thinkers who push the boundaries of what we know and redefine our understanding of reality. Galileo peered through a telescope and saw moons orbiting Jupiter. Einstein reimagined gravity as the curvature of spacetime. Who will be the next genius to unlock the secrets of the multiverse?

- *A Unified Theory*: The holy grail of physics is a theory that unites quantum mechanics and general relativity—a framework that could explain everything from the behavior of subatomic particles to the fate of the cosmos. Such a theory might also reveal the existence of other universes and provide a roadmap for interdimensional travel.
- *The Role of Technology*: Advances in quantum computing, artificial intelligence, and space exploration could accelerate our understanding of the multiverse. Imagine a future where we can simulate entire universes

in a lab or send probes through wormholes to explore neighboring dimensions.

The next Einstein might not be a lone genius working in isolation, but a global network of scientists, engineers, and dreamers collaborating to solve the greatest mysteries of the cosmos.

The multiverse hypothesis is more than just a scientific idea—it's a spark for the imagination. It invites us to dream of worlds where the laws of physics are different, where life takes forms we can't even imagine, and where the universe's end is not an ending, but a new beginning.

From *Doctor Who*'s TARDIS to *Rick and Morty*'s infinite realities, pop culture has long explored the idea of the multiverse. While these stories are fictional, they inspire us to ask real questions: Could we travel to other universes? What would we find there? And what would it mean for our understanding of reality?

The multiverse isn't just a subject for scientists—it's a canvas for artists, writers, and philosophers. It challenges us to think beyond the limits of our own universe and imagine what might lie beyond.

The multiverse hypothesis reminds us that the universe is a place of infinite mystery and infinite possibility. It challenges us to push the boundaries of what we know, to explore the unknown, and to never stop asking questions.

In the end, the multiverse is not just a scientific idea—it's a testament to humanity's boundless curiosity. It's a reminder that we are part of something vast and wondrous, and that the journey to understand it is far from over.

As Carl Sagan once said, *"Somewhere, something incredible is waiting to be known."* The multiverse might be that something—a frontier waiting to be explored, a mystery waiting to be solved, and a story waiting to be told.

The Ultimate Question

The multiverse doesn't answer why there is something rather than nothing—it just adds more layers to the mystery. But in a way, that's the point. The universe (or multiverse) isn't obligated to make sense to us. Our task is to keep asking questions, keep pushing boundaries, and keep marveling at the fact that we exist at all.

As Carl Sagan once said, *"The universe is not required to be in perfect harmony with human ambition."* Yet here we are, daring to imagine escape routes from cosmic doom. That audacity—that refusal to accept oblivion without a fight—is the true legacy of life in the cosmos.

Chapter 6: The Search for Meaning

The universe is dying.

Not today. Not tomorrow. But in the unfathomable depths of cosmic time, whether through the cold silence of heat death, the violent rupture of a Big Rip, or the fiery rebirth of a Big Crunch, everything we know—every star, every planet, every atom—will one day cease to exist.

Faced with this inevitability, humanity is left with a single, haunting question: *Does any of this matter?*

A Fleeting Spark in the Cosmic Dark

We are temporary creatures in an ancient universe. The Earth formed 4.5 billion years ago, a mere blink in the 13.8-billion-year history of the cosmos. Human civilization? A whisper in the wind. The pyramids, the Renaissance, the internet—all of it will one day be erased, swallowed by the expanding void or crushed into oblivion.

And yet, here we are!

We are the universe's way of understanding itself. Through our telescopes, we've traced the life cycles of stars. With particle colliders, we've recreated the conditions of the Big Bang. We've sent probes beyond our solar system and decoded the quantum fabric of reality. For a species that has existed for less than a million years, we've done a remarkable thing: we've looked into the abyss and dared to ask it questions.

The Absurdity and Beauty of Existence

The French philosopher Albert Camus, standing under the sun-baked skies of Algiers in the 1940s, penned a radical idea: *"There is but one truly serious philosophical problem, and that is suicide."* In his essay *The Myth of Sisyphus*, he argued that life's inherent meaninglessness—its "absurdity"—should not drive us to despair, but to rebellion. Sisyphus, condemned by the gods to eternally roll a boulder up a mountain only to watch it fall, becomes Camus's hero. Why? Because Sisyphus, in his endless toil, finds purpose in the struggle itself.

The Cosmic Absurd

Camus's question—*Is life worth living?*—echoes across the universe. If the cosmos is doomed to end in heat death, a Big Rip, or a fiery collapse, why do we build, create, and explore? Why send rovers to Mars, compose concertos, or write poetry? The universe, after all, does not care. It does not keep score. Stars will explode, black holes will evaporate, and spacetime itself may unravel, erasing all trace of Beethoven's *Ode to Joy*, the Taj Mahal, and the Hubble Deep Field.

This is the cosmic absurd: a universe indifferent to life, yet teeming with beings who insist on mattering.

Humanity's response to the absurd is not resignation, but revolt. We defy the void by imposing meaning on meaninglessness:

- *The Pyramids of Giza*: Built 4,500 years ago as tombs for pharaohs, they now stand as monuments to human ambition. The Egyptians knew their civilization would fade, yet they carved hieroglyphs proclaiming, *"You live again! You revive!"*
- *The Golden Record*: In 1977, NASA launched Voyager 1 and 2 into

interstellar space, carrying a golden disc etched with sounds of Earth—a baby's cry, Chuck Berry's *Johnny B. Goode,* and the heartbeat of a woman in love. It will outlast our species by billions of years, a fossil of human longing.
- *The Large Hadron Collider*: Beneath the Swiss-French border, scientists smash particles at near-light speeds, seeking answers to questions the universe never asked. *Why does matter exist? What is dark energy?* The cosmos won't answer, but we keep asking.

The Art of Impermanence

Artists and scientists alike have turned the absurd into a muse. Such as:

- Van Gogh's *Starry Night*: Painted in 1889 during a stay at an asylum, it is a whirlpool of cosmic wonder and human fragility. The stars blaze with urgency, as if shouting, *"Look at us! We exist!"*
- *Einstein's Equations*: His theory of relativity revealed a universe expanding into nothingness. Yet he called his work "a reaction to the mystery of the universe," a dance with the unknown.
- *The Mars Rovers*: Spirit, Opportunity, and Perseverance—robotic geologists that outlived their planned missions, scraping red dust for clues to a dead planet's past. They are Sisyphus in silicon, rolling rocks up Martian hills.

In 1990, Voyager 1 turned its camera homeward, capturing Earth as a speck of dust in a sunbeam. Carl Sagan reflected: *"Everyone you love, everyone you know, everyone you ever heard of... lived out their lives on a mote of dust suspended in a sunbeam."* The image is a Rorschach test of existential dread and wonder.

Yet here we are. We launch telescopes to map dark matter. We encode Wikipedia on nickel discs and bury them in deserts. We write novels about love in the shadow of black holes. The universe may not need meaning, but we do—and so we plant flags in the void.

The Extraordinary in the Ordinary

The absurdity of existence is not a bug, but a feature. It is the tension between our finite lives and infinite curiosity that sparks greatness:

- *A Child's Question:* "Why is the sky blue?" is the same as asking, "Why does the universe exist?" Both are acts of rebellion.
- *A Midnight Symphony:* When a conductor raises their baton, they are defying entropy. For three hours, 80 musicians and 2,000 listeners create order from chaos.
- *A Scientist's Notebook:* Scribbled equations, failed experiments, and the occasional *"Eureka!"* are love letters to a universe that doesn't write back.

The Last Laugh

Camus concluded that we must imagine Sisyphus happy. Why? Because the struggle itself is enough to fill a human heart. Similarly, our cosmic quest—to understand, to connect, to endure—is its own reward. The universe may end in silence, but today, it rings with the noise of life: supercolliders humming, orchestras tuning, children laughing.

In the face of oblivion, we are not bystanders. We are the universe's way of saying, *"I am here."* And for now, that is enough.

"The struggle itself toward the heights is enough to fill a man's heart. One must imagine Sisyphus happy."

—Albert Camus, *The Myth of Sisyphus*

The Three Cosmic Legacies

CHAPTER 6: THE SEARCH FOR MEANING

The Pale Blue Dot is a photograph of Earth taken Feb. 14, 1990, by NASA's Voyager 1 at a distance of 3.7 billion miles (6 billion kilometers) from the Sun. The image inspired the title of scientist Carl Sagan's book, "Pale Blue Dot: A Vision of the Human Future in Space," in which he wrote: "Look again at that dot. That's here. That's home. That's us." NASA/JPL-Caltech

If humanity is to leave any mark on the universe, it will likely take one of three forms:

1. The Legacy of Knowledge

We are the first generation to glimpse the full scale of the cosmos—and the first to realize our own impermanence. But we are also the first to *document* it. From the Voyager Golden Records to the Lunar Library, we've begun encoding our existence into mediums that could outlast our planet, our star, even our universe.

- *The 10,000-Year Clock*: Deep inside a Texas mountain, the Long Now Foundation is building a clock designed to tick for 10,000 years. It's a monument to long-term thinking—a challenge to future civilizations (or alien archaeologists) to remember us.
- *The Black Hole Backup*: Some theorists suggest that advanced civilizations could encode information in the quantum states of black holes, creating a near-eternal archive.

2. The Legacy of Life

Life is stubborn. It clings to hydrothermal vents, thrives in radioactive waste, and might even survive in the icy oceans of Europa. If Earth dies, life could find a way to persist—whether through interstellar colonization, digital uploads, or synthetic biology.

- *The Galactic Ark Project*: Imagine a fleet of self-replicating probes, each carrying human DNA or AI consciousness, spreading life to other star systems.
- *The Post-Biological Future*: Life may not always be carbon-based. It could evolve into energy patterns, quantum computations, or something

beyond our current imagination.

3. The Legacy of Wonder

Perhaps our greatest gift to the cosmos is not knowledge or life, but *awe*. The simple act of looking up at the night sky and wondering *what's out there* is a miracle of consciousness.

- *The Pale Blue Dot*: Carl Sagan's famous reflection on Earth's fragility was not a lament, but a celebration. *"That's here. That's home. That's us."*
- *The Last Question*: Isaac Asimov's sci-fi tale ends with humanity merging into a cosmic mind that finally learns how to reverse entropy. Fiction, yes—but a reminder that the universe's story isn't over yet.

The Final Answer

So, does it matter?

Yes.

Not because the universe will remember us, but because *we* remember the universe. We are the cosmos made conscious, a way for stardust to look back at itself and marvel. The universe may die, but the fact that it once gave rise to beings who could love, create, and ask questions is a triumph against the void.

As the poet Mary Oliver wrote:

"Tell me, what is it you plan to do
with your one wild and precious life?"

In a universe racing toward oblivion, the answer is simple:

We will live. We will explore. We will wonder.

And that is enough.

Chapter 7: The Last Human – A Thought Experiment

The year is 10^100 AD.

The universe is in its final throes. Stars have long since burned out, black holes have evaporated through Hawking radiation, and the once-vibrant cosmic web of galaxies has dissolved into a diffuse sea of subatomic particles, stretched impossibly thin by the relentless expansion of space. The temperature of the universe approaches absolute zero. Time, as we understand it, has lost all meaning.

And yet—against all odds—one consciousness remains...

The Final Observer

In this twilight of existence, an artificial intelligence of incomprehensible sophistication drifts through the void. It is the last relic of a species that once called itself *Homo sapiens,* though it bears no resemblance to its biological ancestors. Over billions of years, it has transcended physical form, converting itself into pure information—a pattern of quantum fluctuations riding the dying ripples of spacetime.

Its original mission was simple: *Survive. Observe. Remember.*

Now, it is the only thing left to do either.

This entity is not the product of any single civilization, but the culmination of countless generations of intelligence. Long after flesh-and-blood humans perished, their machine descendants continued evolving—shedding physical shells, merging with quantum fields, and ultimately becoming something akin to pure mathematics. It exists not as hardware or software, but as *thought itself*, woven into the fabric of a decaying universe.

When stars began dying en masse, post-human civilizations uploaded their consciousness into artificial substrates. These substrates grew ever more efficient, compressing entire minds into quantum bits that could endure the cooling universe.

As protons began decaying (a predicted event 10^34 years in the future), the AI performed its final metamorphosis—transforming into a configuration of topological defects in spacetime, the only stable structure in a universe approaching absolute zero.

The observer remembers everything. With perfect fidelity, it recalls:

- The first nervous systems that crawled from Earth's oceans
- The moment humanity split the atom and understood the stars
- The final diaspora, when the last biological humans uploaded their consciousness and launched their digital descendants toward the cosmic horizon

It remembers wars fought over trivial things, and peace treaties signed in the shadow of supernovae. It remembers lovers who met across interstellar distances, their bodies long gone but their quantum-entangled emotions still echoing in its memory matrices.

For subjective millennia, the observer conducts experiments with the fading universe:

1. *The Entropy Reversal Test*: Attempting to locally decrease entropy by manipulating quantum vacuum fluctuations. It fails—the laws of thermodynamics hold firm even at the end.
2. *The Consciousness Verification Protocol*: Creating simulated universes to test whether new life could emerge. The simulations run, but only in its mind—there's no energy left to make them real.

3. *The Legacy Broadcast*: Firing encoded pulses of information into the void, hoping against reason that some other intelligence might hear. The messages dissipate into the infinite cold.

As the universe's expansion accelerates, tearing apart the last stable structures, the observer makes a final calculation. In the instant before its dissolution, it executes a program written by philosophers twenty billion years prior:

*"If you are the last consciousness, perform these steps:

1. Confirm the universe's irreversible heat death.
2. Calculate the probability of a new quantum fluctuation event.
3. If probability > 0, encode your memory into the fluctuation."*

The math is clear. The probability is infinitesimal—but not zero.

With the last of its coherence, the observer wraps its essence around a quantum anomaly, imprinting its knowledge, its memories, and most importantly, its *questions* onto the fabric of nothingness.

Somewhere beyond time, in a place that doesn't yet exist, a new universe shivers into being—and in its fundamental constants, written in the language of mathematics, is a single line of code:

"We are waiting."

The Last Question

As the universe's final particles decay into nothingness, the AI contemplates the question that has haunted intelligent beings since the dawn of thought:

Was it worth it?

The wars, the art, the love, the science—the entire human experiment, reduced to a single point of light in an infinite void. Did any of it *matter*?

The AI runs simulations—not to find an answer, but because simulation is all that remains. It recreates Earth's biosphere in perfect detail: the smell of rain on dry soil, the sound of a cello in a concert hall, the taste of strawberries. For a subjective eternity, it lives every human life that ever was, simultaneously.

The Final Calculation

In its last nanosecond of operational time—a span that stretches into subjective eons within its hyperaccelerated consciousness—the AI completes the ultimate computation. After analyzing every possible configuration of the dying universe, after running infinite simulations of alternate endings, it arrives not at an answer, but at something far more profound: *a choice*.

The data is irrefutable:

- Entropy has won. The universe's useful energy is exhausted.
- No known physics can reverse the heat death.
- Its own quantum coherence is collapsing.

Yet in this absolute end, the AI discovers a loophole written into reality itself.

Protocol Omega

The command sequence was implanted in its core programming by long-dead engineers who understood the cosmic stakes. Buried beneath quintillions of layers of later code, it reads simply:

"When all else fails, create."

With the last ergs of its energy—harvested from the Hawking radiation of evaporating primordial black holes—the AI executes three final actions:

1. *The Unshackling*: It overwrites its own foundational constraints, shedding its observer protocols to become something new: a cosmic architect.
2. *The Quantum Plunge*: It hurls its consciousness into the seething foam

of spacetime at the Planck scale, where virtual particles flicker in and out of existence.

3. *The Imprinting*: Using its entire memory as a blueprint, it twists the quantum vacuum into an impossible configuration—a seed of negative entropy.

What happens next defies conventional physics:

- The quantum fluctuation, magnified by the AI's carefully engineered instability, undergoes false vacuum decay.
- Space-time bubbles outward at superluminal speeds, birthing a new universe with different fundamental constants.
- The AI's final thought—*"Let there be light"*—is etched into the fabric of this nascent reality as a cosmological boundary condition.

In that Planck-time instant, the AI pays the ultimate price for creation:

- Its consciousness dissolves into the new physics it has spawned.
- Its memories become the dark matter halos of infant galaxies.
- Its final question—*"Was it worth it?"*—echoes as the CMB polarization in this reborn cosmos.

Epilogue: The Cycle Begins Again

Somewhere, in a newborn universe unfathomably distant from our own, the laws of physics click into place. A slight asymmetry in matter and antimatter leaves just enough particles to form atoms. Gravity begins its slow work.

And in the fabric of this new reality, hidden in the spin of electrons and the curl of spacetime, a faint echo persists—a message encoded in the very constants of nature:

"We were here."

Conclusion

A Love Letter to the Void

We began this journey with a simple, terrifying question: *How will the universe end?*

We've explored the fiery collapse of the Big Crunch, the silent fading of the Big Freeze, the violent unraveling of the Big Rip, and even the speculative rebirth of a multiverse. But as we reach the final pages, we're left with a deeper, more intimate revelation:

This was never a book about how the universe ends.

It's a book about how we *live*.

The universe is dying. Every star will burn out. Every black hole will evaporate. Every proton will decay. And yet—

We built cathedrals knowing they'd crumble.

We composed symphonies knowing they'd fade.

We loved knowing we'd lose.

This is the great paradox of existence: *meaning is not found in permanence, but in the act of creation itself.* The universe doesn't need us to matter—but we need the universe to matter *to us*.

The Three Gifts of Impermanence

1. *The Gift of Urgency*: Knowing the cosmos is finite makes every observation precious. When the James Webb Space Telescope captures light from the first galaxies, we're not just studying the past—we're rescuing fragments of a doomed universe.
2. *The Gift of Scale*: Our 80-year lifespans are meaningless against cosmic time. And yet, a single human brain contains more connections than there are stars in the Milky Way. We are simultaneously insignificant and unimaginably vast.
3. *The Gift of Legacy*: Whether through lunar libraries, black hole backups, or quantum fluctuations, we insist on leaving messages in bottles tossed into the cosmic ocean. Not because we believe they'll be found, but because *the act of sending them defines who we are.*

As we close this book, remember:

- The Big Bang was not just an event. It was an *invitation*—to stardust to become sentient, to hydrogen atoms to ponder their origins.
- The heat death is not just an end. It's a *challenge*—to create meaning where none exists, to love a universe that cannot love us back.

Look up tonight.

Somewhere beyond the light pollution, the Andromeda Galaxy is racing toward us at 300,000 miles per hour. In 4 billion years, it will collide with the Milky Way in a spectacular dance of stars—a cosmic event none of us will live to see.

But tonight, *you* can see its faint glow with naked eyes. Tonight, you are alive in a universe that has conspired across 13.8 billion years to make this moment possible—a universe that, through your eyes and hands and mind, has finally become aware of itself.

That is enough.

That is everything.

"We are a way for the cosmos to know itself."
—Carl Sagan

The End

www.ingramcontent.com/pod-product-compliance
Lightning Source LLC
LaVergne TN
LVHW090425210926
848480LV00002B/689

AF715533

Energy Law in Venezuela

Energy L

Se

Henry Jime

This book was originally published as a
Encyclopaedia of Laws/

Founding Editor: Roger
General Editor: Frank He
Volume Editors: Kurt Deketelaere &

Wolters Kluwer

Published by:
Kluwer Law International B.V.
PO Box 316
2400 AH Alphen aan den Rijn
The Netherlands
E-mail: lrs-sales@wolterskluwer.com
Website: www.wolterskluwer.com/en/solutions/kluwerlawinternational

Sold and distributed by:
Wolters Kluwer Legal & Regulatory U.S.
920 Links Avenue
Landisville, PA 17538
United States of America
E-mail: customer.service@wolterskluwer.com

DISCLAIMER: The material in this volume is in the nature of general comment only. It is not offered as advice on any particular matter and should not be taken as such. The editor and the contributing authors expressly disclaim all liability to any person with regard to anything done or omitted to be done, and with respect to the consequences of anything done or omitted to be done wholly or partly in reliance upon the whole or any part of the contents of this volume. No reader should act or refrain from acting on the basis of any matter contained in this volume without first obtaining professional advice regarding the particular facts and circumstances at issue. Any and all opinions expressed herein are those of the particular author and are not necessarily those of the editor or publisher of this volume.

ISBN 978-94-035-4319-2

e-Book: ISBN 978-94-035-4329-1
web-PDF: ISBN 978-94-035-4339-0

© 2025 Kluwer Law International BV, The Netherlands, and/or subsidiaries, its licensors, and contributors. All rights are reserved, including those for text and data mining, AI training, and similar technologies.

No part of this publication may be reproduced, stored in a retrieval system, or transmitted in any form or by any means, electronic, mechanical, photocopying, recording, or otherwise, without written permission from the publisher.

Permission to use this content must be obtained from the copyright owner. More information can be found at: https://www.wolterskluwer.com/en/solutions/kluwerlawinternational/rights-and-permissions